George Sandeman

Problems of biology

George Sandeman

Problems of biology

ISBN/EAN: 9783337215057

Printed in Europe, USA, Canada, Australia, Japan

Cover: Foto ©berggeist007 / pixelio.de

More available books at **www.hansebooks.com**

PROBLEMS OF BIOLOGY

BY

GEORGE SANDEMAN, M.A.

LONDON
SWAN SONNENSCHEIN & CO., LIMITED
PATERNOSTER SQUARE
1896

PREFACE

THIS volume contains the criticism of the contemporary biological systems. That enquiry is necessary as an introduction to the study of the problems of organic life, but it is not in itself a doctrine of biology. The argument ought to proceed to the discussion of the philosophy of nature.

I have to thank Dr. C. M. Douglas, Dr. George R. Wilson, and Mr. A. B. Macaulay, for suggestions in connection with various parts of the book.

G. S.

COLINTON, *October*, 1896.

CONTENTS

	PAGE
I. METHODS OF BIOLOGY	1
II. THE FIRST POSTULATE OF BIOLOGY	46
III. THE SECOND POSTULATE OF BIOLOGY	89
IV. THE THIRD POSTULATE OF BIOLOGY	139
V. THE UNITY OF THE ORGANISM	196

PROBLEMS OF BIOLOGY

CHAPTER I.

METHODS OF BIOLOGY.

THE aim of this essay is to inquire into the nature of the problems which come before us in the study of animals and plants, in order that we may define the right method and the limits of biology, and that we may obtain some adequate criterion by which to judge of the merits of the numerous theories which have arisen in explanation of facts about organisms. Its method is throughout that of inquiry; and I shall build up no systematic hypothesis as a solution of the problems. For we shall find that such hypotheses, as are most commonly used in this matter, bring with them certain unavoidable disadvantages, in that they are themselves in contradiction with the very possibility of research, and serve rather to hide the problems than to give any satisfactory account of them. And our material is two-fold; for there are the theories of biology on the one hand, and on the other the records of research; and these two sources are, for the critical study of problems, equally important.

Every serious student of biology must be conscious of a remarkable anarchy within the science, as well as of a certain indefiniteness in its scope. These inner and outer confusions, together with the complex and uncriticised meanings of biological terms, supply reasons why we should supplement and interpret biological theory by a critical method; and that method seems to be equally necessary to any clear understanding of the problems, whether the conclusions of the present essay are found to be true or abstract. I shall describe the confusions in question, leaving the discussion of the reasons for them to a later place.

When the methods and form of any science are discussed, the argument moves between two dangers. For it is possible to lay down very close limits for the content of the science, and to define its methods with a hard pedantry, so that the conclusions which are based on such limitations are only true of a part, and that, perhaps, not an essential part, of the science. And, on the other hand, one may lightly enumerate the more obvious points of its content, and decorate the discussion with many apt illustrations of its method, yet gain no distinct impression of the unique business of the science. On the whole, I shall attempt to err in neither of these directions. For the purposes of this essay, biology includes the systems of explanation of the forms and functions and origins of animals and plants in general. For us, it is to be a science which is definitely distinct from morphology, physiology, embryology, and the other studies which deal with aspects, parts, and periods of organisms. Biology is not to be a mere encyclopædia of facts about organisms, it is in some way to include, and yet to be distinct from those other studies. It is not the science of form or of function, because it is the science of the union of those aspects, and of the union of morphology and physiology with the studies of the developing individual and the developing race. And since there is no need of a science to include the more special sciences merely side by side, the uniting action of biology is not the mere addition of one of them to the others. However narrowly we define the science for our present purpose, it will still include all that is commonly known as biology. We have to do with theories of the development of the individual, of inheritance, and of adaptation, such as those of Lamarck, Darwin, Naegeli, Weismann, and many other authors.

Now the inner confusion of biology depends on the form

of the science; not on its necessary form, nor on its professed form, but on its actual form, which is the strange result of the other two. The necessary form is, as we shall see, a theory of individuality. The professed form is that of the induction of general laws from known facts, with the interpolation, here and there, of hypothetical facts which are supposed to be in the same kind, to fill the lacunæ which are unavoidably left by tardy research or defective instruments. The actual form is a curious product of these two factors, and its several features are referable to their cooperation. For the great body of biological literature, as distinguished from the archives of research which have no theory behind them, is given to us in the form of a large number of independent systems; and the quality of that independence is the source of the inner confusion. Speaking generally, one may say that the systems are each of them divided into two parts; each includes a theory of the development of the individual and a theory of the progressive adaptation of the race; that is, a theory of ontogeny and another of the factors of organic evolution. Or in other words, each biological system has to answer two questions—How are the qualities of the individual related to one another? And—How do the qualities exist by reason of their significance? Every system, at least, which has achieved any fame, has given a complete account of these two problems.

The systems, therefore, deal with the same matter, and the same wealth of recorded research is before them. As sober inductions built upon the sure ground of observation, they should, one would judge, differ from one another only in unimportant points. Indeed, they should so pass into one another that the character of independence of systems should be impossible to them. They should form one body of knowledge, going ever hand in hand with research, informing its method, and certainly capable of being affirmed or denied by a final reference to fact. There should be no such opposition and bare exclusion of one another as that to which we are accustomed.

But consider their actual form. Each of them is complete and final, and will explain, in its own manner, everything which is organic. Each stands over against the others in the form of bare exclusion, and is hard and polished in its independent perfection. And judgment cannot be made between two such mutually contradictory systems by the most exhaustive and ingenious test of facts. There is a complete independence of one another, and an almost complete independence of research. If it were not so, then they would combine, and research would discriminate between them. There are twenty good theories of the development of the individual, but I cannot say that any one seems to be better or worse than all the rest. A certain controversy with regard to natural selection and use inheritance lived long, and was discussed in every public place and with the aid of hosts of detailed observations. Yet it was never cleared up, and neither side had the advantage; but because men became weary of it, it has now been allowed to rest. It is not otherwise with the history of biology. New systems supersede old ones, and the latter are not disproved but forgotten. If this seems to be a too sweeping judgment, then the archives of research are before you, and the theories are elaborately stated for every detail and exhaustively illustrated from facts about organisms; will you tell—not what maintains the form of the individual from day to day, for the theories abstract from that consideration—but what determines the rise of that individual form; whether it be the clairvoyance of the unconscious, the immanent soul, the gemmules, the idioplasm, the pangenes, the stirp, the physiological units, or the struggle of parts; that the palm may be given to Hartmann, Stahl, Darwin, Weismann and Naegeli, de Vries, Galton, Spencer, or Roux? Or perhaps some other hypothetical agent seems to you preferable? For my part, if the method of any one of these theories be allowed for a moment, I cannot see any reason for either affirming or denying any one of them. And if one of these principles made my hand, I have no doubt that it made my foot also, and I cannot be sure that it did not make my argument.

But it may be said that after all we have to some extent moved in accordance with research. Bonnet's belief that the image of the adult was contained in the germ in little, difference for difference, has been expelled by the progress of microscopic investigation. Lamarck's *cause excitatrice* has disappeared before physiological research. And on the other hand, Weismann's idioplasm does correspond to granules which stain with logwood, and the struggle and selection of parts is certainly a fact so long as the parts are in any way related to one another. That is merely to say that hypothetical agents and hypothetical processes disappear before research; and that the theories invent new processes and new determinant agents in as yet unexplored forms. We might point out that the agents which are at present in vogue are, in fact, unknown, but it is better to await the more conclusive demonstration that they are in necessity unknowable. These are some of the features of the inner confusion. They have made the very name of biology a by-word. And though the anarchy may not be obvious to a people delighting in formulæ which may be applied with equal facility and barrenness to everything which is organic, it is so present to men of research that they leave the whole matter on one side as simply not pertinent to their occupation, and are not patient to bear even the mention of what they repudiate, with more justice than they are always aware, as metaphysics.

The outer confusion is not less notable. I have indicated that the formality and abstractness of biological systems is such that they are not affected by the differences in their content; and that a particular fact has not its particular significance to them, but only its general significance as a fact about an organism. But we shall find a graver sign of that formality in the fitness of the systematic formulæ to explain facts which do not come under the special sciences of living things, but which still are facts about individuals or systems. It is not necessary to detail the particular results of the outer confusion. We have seen the arts of human life ascribed to the factors of organic evolution, and a vast literature has by this time accumulated under the form that every evolution, that is, everything individual, is a biological evolution, that is, is organism. I would be slow to condemn any wide view of the organism on account of its width, but danger arises when distinctively human qualities become mere adjectives of protoplasm. It is probable that the popular quasi-biological evolutionism is yet the promise of that working doctrine of individuality for science which at present we cannot find; but so long as the biology from which it draws its eager metaphors is nothing but the denial of individuality, the formal application of the logic of bare identity to the organism, we can expect no stable construction from such theories. In any case, the authority of biology over individuals of every kind which are not organisms, and the degree of its authority over the qualities of men, remain at present undefined; and such indefiniteness is due to the abstractness and the fallacy of that method which can so freely be applied to what you please.

We must include among the features of biology which demand criticism, the vagueness, ambiguity, or positive self-contradiction of its most important terms and conceptions; for this character will also be reduced to a common origin with the others. What, for instance, is *function?* The term is used as undefined and as needing no definition because of its simplicity. If it is all the qualities of a given part, then what is a *functionless structure?* If it is the *end* of a part, that respect in which it subserves the whole, then what are its other activities, and in what respect are they not its *end?* We learn that structure precedes function in ontogeny, and, for the purposes of abstract embryology, we know what the phrase means. But the thesis is a biological thesis, and biology deals throughout with the relation of structure to function *in general.* And I cannot think that such a general consideration can be firm while its terms are capable of the most various meanings. If the structure of to-day answers to the function of to-morrow and not of to-day, then it is not to-day's structure. Such an objection can only seem captious until one has learned from the theories that the vehicle of qualities makes to-day's structure and to-morrow's function, and that *therefore* they fit one another. Such confusions of terms are the very source of controversy, for the latter never lives except upon the platform of a common and uncriticised fallacy. Are acquired qualities inherited? Tell me first of a quality which is not acquired, or of one which is not inherited. Is variation purposive or no? Let us first find anything organic which is not purposive. I do not wish to prejudge such matters; but it certainly seems necessary that they should be judged.

There are thus certain reasons why we should examine biological methods. There are also certain apologies which may be made for the method of criticism which dominates this essay; and they must be mentioned in this place because of the lively objections which that method must call forth in the reader's mind. Criticism is not denial, and an interpretation and comparison of systematic theories is only possible by means of the demonstration of the various planes of abstraction on which they treat the problems. In thus bringing them together and studying them in relation to research, we come really to understand them for the first time. For what is explicit in a theory, is in most cases by no means its most important part; and the postulates which have been taken for granted, the infallible method of observation, and the ambiguities of conceptions, contain what is most influential in the systems. When their point of view is given, the rest follows of necessity ; but it is the business of the student of theory to define the point of view. Further, this method of bringing implicit foundations into the light is the only means we have for arriving at a statement of problems, for such a statement does not form part of the programme of a systematic hypothesis. For the hypothetical agent, by means of its self-contradiction, assumes the whole problem into itself, and appears to be adequate to its solution only so long as that self-contradiction is not made manifest. In this way, as we shall see, all the various forms of vehicles of qualities hide, but cannot be said to be aware of the problem of the nature of organic individuality. In like manner the descriptions of the hypothetical processes of organic evolution hide, but are never the means of stating, the problem of the nature of organic qualities. For the same reason the critical method is necessary to biology, if that science is to present any continuity in its development, and any germinal relationship to research, because we must expose an implicit thread of continuity in the systems, and discover the manner in which the theories abstract from fact in order that they may take the form of systems. It would be a hopeless task to examine all the theories in detail, and to compare them externally. That has often enough been done with but little result, for this one reason, that their implications were not made evident.

But a more important defence for the method is its success in the interpretation of organisms and of the methods of describing them. The history of biology shows that critical doctrines have had their place in the development of the science, and that their influence has been greater and more permanent than that of the hypothetical theories. The doctrine of type, for instance, as followed by such men as Goethe, Geoffroy St. Hilaire, and Dugès, as well as by more recent authors, has been the clearest guide to morphology. It is true that it was merely morphological and abstracted entirely from process and change; but that is equally true of the works of Darwin, Naegeli, de Vries, and Weismann. It differed from the theories of these workers, however, in that it was an incontrovertible presentation of the nature of morphological qualities. For our recent schools deny, in their first postulate, the unity of plan in the structure of the individual, not recognising the necessity of organic proportions. The doctrine of vitalism, again, may be called critical, in that it does no more than to propound the problem of organic activity. Changes, it says, are conditioned in every case by *ends;* as the doctrine of type asserts that measurements are in every case conditioned by proportions. These two doctrines differ from the hypothetical biology, in that in their case room is left for the progressive discovery of the *how* from observation, and that the teleological problem remains in evidence. That the *how* is abstracted from by the hypotheses may be easily learned from their treatment of inheritance. For, in that respect, as in others, they take no account of real process, yet supply its place with impossible and unknowable process. They take no cognisance of the rights of physiology, and, more fatally, contradict its possibility with their quasi-physiology. For the matter is treated as though variation were variation of the adult image —this quality having appeared, and that other quality having disappeared—whereas variation must be variation of the germ, or, better, of the whole. In the same way we must not think of inheritance as immediate, but as mediate; and the middle term is not—except for hypothesis—the abstract qualities of the adult image carried each in its respective vehicle. A special case of the same abstraction is the habit of regarding structure as maintaining itself from day to day by no reason at all, as though it were made of cast iron, and persisted in a material which had no interest in the pattern. Yet this body, incomparably stable as it is, in the midst of ever varying accident, is maintained daily in material which is incomparably unstable—surely this occurs *somehow*. The

Protozoa have complex structures and well-ordered changes, yet many of them can put on form, and cast it off like a garment. And if the monad has its lash mediately because of evolutionary processes which have gone on through a very great number of years, yet it has it also immediately because of something else, for the lash has gone and has appeared again before your eyes. That other *because* has its rights—even prior rights; for that other *because* is what is inherited. Yet it is not even formally allowed for by the hypothetical systems. It is not necessary to multiply examples of the abstractions by which alone the hypotheses can work. But it is largely in respect of such dogmatic abstractions that I would negatively characterise the critical method.

But the most important contributions under this method are those which we owe to the professed students of philosophy, and the last part of my apology must deal with the right to allude in parts to their work about organisms, as to certain biological doctrines among others, in our study of biological theory in general. I am aware of the maxims about the rights of every science in its own sphere. So far as the sciences of inorganic nature, and the abstract sciences of organisms are concerned, those statements have a certain though a limited validity. But the doctrine of the independence of science from philosophy, always over emphasised, has, in the case of biology, no meaning whatever. In a word, the problem of philosophy as regards organisms is the problem of biology; and you may regard either development you please as an aberrant form of the other; for the present point is merely that they are the same. The most concrete science of life may, for our purpose, be called by any convenient name. But we can, at least, be sure of its content. It deals, in alternative ways which absolutely exclude one another, with the ideal unity which underlies organic differences in the individual; it effects a synthesis of abstract sciences by means of a theory of individuality; it brings nature under the form of self-consciousness. On that ground we may regard the critical study of organisms *as organisms* as having the right to a place among the hypothetical theories of organisms *as bare organisms*, and we may justly use it as their criterion.

The theories which we shall examine will be found to deal with *one* feature of organisms alone, and that feature is *universal.* . That is to say, it belongs to all organisms because they are organisms, and to every part and quality and stage of organisms. We judge that the feature is *one* because the theories exclude one another; and we judge that it is *universal*, because the particular differences of organisms do not affect it, so that it is impossible to conceive of any organic forms or functions which would not support the formulæ of the theories, so long as they were the forms and the functions of organisms or of their parts. It is universal because the hypotheses are empty of differentiated content and formal, and because to them, every phenomenon is of equal and of one significance. Now this universal feature in organisms is represented in the systems by means of the hypothetical agent. One need only mark the diversity of hypotheses in order to see at once that the agent is universally necessary, that it does not signify in the least what form the agent takes so long as it does its work, and that the peculiar nature of its activity is never expressed. We shall find that the agent has many qualities, of which we may here glance at the most significant. It is that in which the possible is actual, it is the vehicle for latent qualities. It is that which persists throughout the manifold changes of the organism, preserving thereby the specific image; and it is that which is one and identical in all parts. It is that reality which lies behind the confusing appearance of the creature, which yet is unknown except through those appearances, and therefore, is mediately and even falsely known. It is the sphere in which qualities are separate, discrete, numerable unities, and is the only avenue of relation between qualities. With regard to the qualities it is purely active, and they to it are purely passive; it is in no relation of reciprocity with them, but is absolutely self determining and other determining. All the *rationale* of qualities is taken up into the agent, which effects this here, and that there, by no reason which can be given. It is the identical substance from which the manifold appearances radiate; it is the secret of their teleological form.

I consider that everyone who marks this half of every system—that half, namely, which deals with the relation of the qualities to one another and to the agent—will be inclined to agree that we have here to do with a properly metaphysical conception. He will recognise in the agent the hypostasis of a logical distinction into a quasi-physiological or quasi-psychological distinction, and will be irresistibly led to compare the abstract identity of the individual organism, whatever form it may take through the caprice of hypothesis, with the *abstract substance* or the *thing in itself* of the popular metaphysic which depends on what is called the *relativist* theory of knowledge. And when he learns that this hypothetical agent is not only necessary to every system of the dogmatic biology, as explaining the development of the individual, but is also necessary to the doctrines of organic evolution which form the other half of every system, he will be willing to reconsider the maxim about the unassailable rights of every science in its own sphere; for this reason, that the hypothetical agent is in every case in contradiction with the very possibility of those researches for which biology undertakes the synthesis.

But one may deny the validity of that maxim on more general grounds than these. Even though the identity of the biological and philosophical problems as regards organisms were not so obvious or so bare as it is, we should still be justified in bringing them together. There are no chasms in science, and so long as knowledge is man's knowledge, it is one. The world is not made in compartments answering to university lectureships.

Certainly if this matter of the independence of sciences had really been dominant in knowledge, we should by this time have had nothing but collectors' catalogues on the one hand, and a dead scholasticism on the other. And many evils do at present come on us from the proverb in question, even when it is not merely a cover for cheap ridicule. Isolation is as abstractly self-assertive as is invasion, and a philosophy which is not in active sympathy with science is making a boast of its inadequacy. And the outstanding evil is that every science suffers from an insular ignorance of what is meant by the others; philosophy, for instance, at present knows little of animals and plants but what it has learned through the biology of hypothesis, and accepts the results of the latter for observation ; and biology is content to find the true differences of organisms in the structure of small particles within them, through an inadequate knowledge of the methods of physics. In a word, one science is only too ready to accept the abstractions of others as answering to the nature of the matter studied. Certainly, on the other hand, there are limits to every man's time. But, as I said, in the case of biology and philosophy, we have other and more definite grounds for denying a division. What are *evolution, epigenesis, individuality, adaptation, quality in general, possibility or latency,* and *necessity?* I consider that these fundamental conceptions of biology, which form the content of its discussions, are conceptions of philosophy also and in the same way.

But biology has been defined in various ways, and these definitions must be compared with its actual endeavour. Perhaps it is most commonly regarded as one among the other sciences of organisms, and as on the same plane with them. In so far as it is not a term roughly applied to the whole study of life collectively, it comes to indicate a sort of supplement to the more definite observational sciences, dealing with distributions, origins, and the like. To such a point of view, biology is a vanguard to the whole force of investigation, dealing provisionally, and therefore more or less hypothetically, with matters which in the end are such as must be studied as form, function, and the like. Now there is a vast deal of inexact information scattered through biological literature, and the theories are on the whole as dependent on it as on research. Still, any one of the theories will be found to treat all morphological and physiological results as merely material for its purpose, and as by no means complete in themselves. To anatomy, structure is ultimate, and has nothing to do with change; but to biology, structure is in some way derivative. And there is no result of any one of its fellow sciences which has not a significance to biology.

Another common definition is even more obviously untrue. It is, that biology is the most abstract of the studies of life, having the object of reducing their results to the form of mechanical and physical processes, and of translating the variety of the qualities of an organism into terms of the physical structure and chemical composition of its protoplasm. I have never seen a more remarkable instance of the self-deceptive power of a pretended method than this. For you will find that *the method* is to be that certain qualities, "primary," are the organism for biology, and that these are the quasi-crystallised manifestation of the fine structure of protoplasm. But *the practice* is to take, for biology, even the beauty and the joy of a creature as on a level with its bones, and to rank its instincts as no lower than the extraordinary variety of objects which are called its protoplasm. On the whole, our science is not a mere chemistry of hypothetical substances, or a mere physic of hypothetical "aggregates."

There is more observation in the definition of biology as the theory of the outer relations of the whole creature. Organism and environment, individual and species, parent and child, are, indeed, relations which are subjects for biology. Communities and colonies, instinct, external appearance, adaptation to circumstances, and other matters relating to the whole organism reappear in every theory. And there is thus a warrant for regarding the *unity of the organism* as the object of biological study as distinguished from those sciences which deal with aspects only. But wholeness is not one aspect among others, but includes all aspects. And, in fact, the systems are ever occupied with all the aspects as well as with the unity, so that biology does not deal with certain facts about organisms, but deals with all the facts in a certain manner. All the facts of morphology, embryology, classification, and the like, are included; but they are included as having a reference to the unity of the organism, and as facts about whole individual organisms. In a great part of its extent, indeed, it deals with facts just as bare facts, without troubling itself as to their concrete variety. It takes organs as organs in general, parts as parts, and periods as periods; and, including them all as *differences*, it studies their manner of relation to one another in the whole organism. In like manner structure, function, the rise of these in ontogeny, and their several normalities and perversions, are all included as aspects, not entirely separate from one another, but of one whole, the organism. So also are habits, surroundings, and advantages, referred by biology to a certain unity. And as these are all qualities, they are brought together by our science as qualities of individuals. Again, the differences of alternative forms, or of successive forms in a cycle of such, are considered in relation to one another, as the members of a unity which, in such cases as these, can have, so far as appearance goes, no *apparent* existence; and the theories are mainly occupied with imagining for it an *apparent, but not yet discovered*, existence. We may, therefore, regard biology as a science which deals with whole organisms, if we qualify that statement by indicating, that as there is no whole organism which is not also its parts and stages and alternative forms, so the science deals with these various differences as they are brought into a certain kind of unity. For the present purpose then, we may define biology as the synthesis of several abstract ancillary sciences of living things, which works by means of a theory of the relation of manifold differences and aspects to one another in the unity of the organism. And we may regard its preliminary task as the formulation of some conception of that unity; for some theory of the nature of an individual and of its qualities is, at least, logically prior to the gathering into one of the various aspects and parts of the organism. In justification of that belief we find a theory of the unity of the organism as the chief element in every system of biology.

It is at this point that we may discriminate between the dogmatic and the critical biology; and the test of the nature of every system is the manner in which it treats of the unity in question. No one can deny its necessity to theory; but it is possible to represent it in two ways. The actual method of the dogmatic biology is, as I said, compounded of its necessary and of its professed form; and of these two the former requires that it should throughout refer all differences to the unity of the organism, and the latter requires that it should take no cognisance of anything which cannot be observed, or which is not at least capable of being seen if our instruments are good enough. For the professed form requires that no distinction should be made, which is not a distinction of one apparent difference, or quality, or part, from another. Now no one has ever seen the unity of the organism lying among its parts. So the dogmatic biology is in the dilemma that the distinction which it must use at every step of its argument does not answer to a phenomenal difference; so that the particular nature of the difference cannot be given, in its theory, from observation of it. Hence its actual form. "The unity," it says, "is *there;* but there can be nothing *there*, but objects of various kinds which are phenomenally different and distinguishable from one another; therefore, the unity is a phenomenal difference of some kind which we have not yet discovered, but which will some day, perhaps to-morrow, be discovered by microscopic research; and in the meantime, since we cannot do without the unity, *I* think it is probably these granules clumped together in the nucleus,[1] but *you* think it is these distributed granules which stain with acid fuchsin,[2] while *he* thinks it is a kind of thinking and willing thing in the various parts which yet, of course, does not think or will in the every-day sense of the word."[3]

Until the unity of the organism has been discovered as a phenomenal difference, it is plain that just these hypotheses as to what kind of difference it is may be multiplied to any extent, and that it is not easy to say to which of them one ought to subscribe. But it is also not less plain that such hypotheses are absolutely necessary to the dogmatic biology. And we have, in fact, very many theories in this kind, some of which will be examined in the third chapter.

Since, according to this method of biology, there is nothing in the organism which is not a difference, and since the unity itself is a difference which externally acts on the others in a quite unique manner, it follows that the differences are not themselves in every case the unity at the same time. They may be called *mere* differences, a conception which in return requires a principle of unity external to them if there is to be unity at all. And on this point the dogmatic biology is not silent. For it, the qualities which make up the organism are independent of one another as separate constituent elements of the body. This point of view arises immediately from the professed form of the science, in conjunction with its necessary form. For the latter requires that biology should in the first place have a theory of qualities as differences and as identity; but observation does not seem to give qualities as identical, but only as this and that mere difference; and these differ-

---

[1] Weismann, etc.   [2] Altmann, etc.   [3] Stahl, etc.

ences are distinguished from one another by observation as one thing from its neighbour. Therefore it says, "Any two qualities are not one thing, but two things; and in the end of analysis the body is made up of a certain number of differences; which, by means of the dominant unity, are kept together and determined separately into their right proportions and successions. A latent quality is just as much one thing as those which are manifest, only that for the time being it is a hidden thing, still preserving its separate existence in another form within the body."

Again, each difference, having its own separate existence, can be changed in any way without affecting the other constituent elements of the organism. It also has it own use or significance, by reason of which it exists; and that use is not found in the other differences, nor is its existence derived from that of the other differences. This is little more than another way of saying what we noticed above, with regard to the independence of qualities. But it indicates that the reason for any quality or part is not to be found in other qualities or parts, but in some external and independent value of its own; and that the existence of one difference is not accounted for or conditioned by the existence of others. In a word, the *rationale* of organic parts is not to be found in their relation to one another, for they have for our theories no such inner relation, but in the external and independent use of each part. And of such qualities, successively added to one another during history (and each persisting in its own right of external usefulness after it has been added), the organism which we see has been put together.

The actual form of the biology of systematic hypothesis is thus dependent on the three postulates which I have indicated, and which are one in origin. They are as follows:—

1. That the qualities of the individual are discrete, numerable, constituent elements of which the organism is

the total sum; and have, therefore, each the value of an ultimate unit for biology. They are thus independent of one another as regards their significance, maintenance, development in the individual, existence when latent, inheritance and variation, and acquirement by the race.

2. That all the qualities of the organism and all its stages are the manifestation of, and are related to one another only through an agent or system of agents within the known body. The agent, which answers to the unity of the organism, is purely self-determining; it is in the attitude of pure activity to the body, which, in consequence, is in the attitude of pure passivity to the agent. The agent may be associated with any particular microscopic appearance in the body, or with any particular states of its feeling, or it may be regarded as not yet discovered; but in every case it is formally allowed to have such a phenomenal existence. It carries the qualities when they are latent and carries alternative qualities, and it manifests these when and where they ought to be manifested. If the agent is a "material vehicle of hereditary qualities" there is a distinct agent to each difference in the organism, but a system may be given to such agents by the hypothesis of a material "architecture" from which they are unfolded. If the agent, on the other hand, be a quasi-psychical principle, there is only one such; and its system of differences, answering to the known differences of the body, is a system of ideas dimly apprehended.

3. That the adaptedness of organisms is due to the external addition of new qualities to the rest, which henceforward are included among, but not conditioned, by the qualities which have, up to that time, existed. The environment is something separate from the organism; and the latter is, by the addition of new qualities to the trust of the agent, thus educated up to circumstances which can exist without it. The inertia of the agent is such that it may persist in presenting qualities which are unrelated to other qualities

and which have ceased to have any special external and independent use. The various qualities of the organism are thus due to the slow addition of modifications through many years of changing circumstances.

Of these postulates, the first is often stated, and often unconsciously present, but it is universally implied in the methods and terms of the theories. The second is a mere generalisation from the theories of ontogeny, which take two forms—that, namely, of the *material vehicle of hereditary qualities* and that of the *quasi-psychical principle*. And the third is the postulate for the theories of adaptation by the transformation of species. The three postulates are not only necessary to one another, but they are also necessary to the theories of the biology of systematic hypothesis. Each of the latter has, therefore, two parts, of which one is an hypothetical doctrine of individuality, and the other an hypothetical doctrine of adaptation.

Now this peculiar method, as revealed in its three postulates, is in many ways in contradiction with the possibility of research. Such contradictions come up repeatedly in the simplest study of the theories, but we need not here notice more than that which exists between the postulated independence of differences and the ordinary methods either of morphology or of physiology. For the proportions of morphology are the denial of the independence of single measurements. And the science of physiology works with the invariably justified assurance that qualities and changes within the individual are causally related in complex mazes of inter-dependence; that is, the differences are known to research under the form of reciprocity. And one can hardly regard that as a germinal form of biology which would make nonsense of morphology and of physiology. Nor is that a stable form of biology which plainly cannot be either refuted or supported by observation.

And we may here repeat what the method really is. It is the hypostasis of logical distinctions into the organism.

Universal and particular become respectively the agent and the body of known differences. Any distinction between differences becomes a chasm in nature, and because qualities are distinguished from one another, they are therefore discrete and numerable constituent elements. Such a method may obviously be defended as a working method; and that defence, if successful, would be justified in using the proverb about the rights of each science in its own sphere. But the defence is not applicable to the present case for two reasons. First, the working hypothesis does not work. It is alone in science, and there is none like it. For one can only know relations, and this is the denial of relations. And nothing whatever can be deduced from it when it has been made, for it is exceedingly abstract and formal. The agent does what we see, but no theorist has ever ventured to predict what it will be found to do in cases not yet observed. It makes my hand, but I cannot judge that *therefore* it will make a foot to answer to the hand. From all its operations of all kinds but one, you cannot show that that other operation is necessary. It is the pure form of arbitrarily doing, *what it does*. And second, the hypothesis is itself only the bare fact of the hypothesis. Qualities are ultimate units, but no two authors will agree as to how the body should be divided, and as to where analysis should stop, to obtain these units. All qualities are the work, not of one another, but of the agent—but we have been given twenty agents of the most various kinds, every one of which is as good as the others. And no hypothesis has yet even barely indicated how the agent does its task, so that the working hypothesis does not even try to work. We have no physiology of the idioplasm or of the immanent soul.

Contrast, on the other hand, the postulates of biology with the atomic theory. The latter provides agents to which the present criticism is inapplicable. For to the one simple form of the reciprocal attractions and impenetrability of atoms the whole diversity of physical facts can be reduced. But there is no simple form for the biological agent; it is more complex than the organism. And reasons can be given for the different behaviours of atoms under different circumstances, but the *idioplasm* is the very principle of unreason, the very denial of the intelligibility of differences in relation to one another. And further, one can hardly call that a working instrument, which the craftsmen do not, in fact, use. Is any one of the theories of agents of any use whatever to the practical embryologist?

These are some of the features of the actual form of biology, and they are at first somewhat difficult to reconcile with the requirements of its professed form. The contradiction of research and the unrestrained hypothesis seem quite at variance with the exhaustive collections of facts which characterise the theories. Can it be possible to build up theories from such innumerable phenomena, without achieving certainty and solidity of construction? The variety of the theories shows the possibility; but the reason of the matter is no less plain. The arrays of facts are in some way rather illustrations for the hypothetical systems, than proofs or supports of them. You cannot criticise theories of this class by referring to different facts, for facts, however numerous and however various they may be, do not discriminate, so long as, for the theory in question, they are facts of the same kind. And, in so far as the systems deal with the nature of individuality and with adaptation and with the nature of qualities, they are dealing with universal features of organisms, and organic differences do not discriminate for them.

You may expect to get light on the *immanent soul* or the *idioplasm*, on natural selection or use inheritance, by the exhaustive enumeration of various individuals and various adaptations, as soon as you learn anything about causality by playing billiards, or throw light on the nature of number by counting things in the street. As numerable, things are similar; and as the adapted qualities of individuals, all the facts about organisms are similar. Everything organic is what it is because of natural selection, in that it has not been killed off; and equally, everything is what it is because of use inheritance, in that it is the product of the whole germ which is the product of the whole parent. In like manner, everything organic belongs to the unity of the organism, and as such will not refuse to admit of the formula either of the idioplasm, or of the immanent soul, or of any other agent you please which represents in all essential respects the unity of the organism.

I said above that the biology of hypothesis is little more than the mere application to organisms of a certain definite metaphysical doctrine, and that this metaphysical doctrine is clearly and fully stated in the three postulates which dominate or rather constitute the theoretic systems in question. Therefore we can only find the thread of continuity in the systems by the examination of this common fallacy which runs through them, but which they do not question. They assume it and the postulates derived from it, as unquestionable, chiefly because the metaphysical dogma which we are about to consider is taken to be the beginning and end of metaphysic, by the teachers of natural science. The latter, indeed, are never weary of stating it, as the sure and ultimate ground of their method, although it is in contradiction with the possibility of any science.

Now the first postulate gave us qualities as independent and discrete constituent elements. And the second gave us the "vehicle of qualities" or agent which effects for them a certain external union. This vehicle was, as it were, the inner side of the quality; it is that in which the true explanation of what appears to us as a mere collocation of qualities is to be found; it is the reality of which the given organism is the mere manifestation. It is the sphere in which possibility and actuality are one; it is unqualified substance or abstract identity. Such characters will be revealed for the "vehicle" in our study of the theories which deal with it. But in this place we shall find the true origin of the independence of qualities as merely collocated things, and of the agent as the "vehicle" of qualities. Our two postulates arise directly from the relativist theory of knowledge, and the agent is the thing-in-itself. In this respect we shall find the biology of hypothesis to be a form of metaphysic.

We may first dispose of an objection which at once arises. It might be said that there is one great difference between the vehicle of qualities and the thing-in-itself, in that the former always receives, formally, a phenomenal existence, while the latter is formally unknowable. But the phenomenal existence of the agent is a mere confusion, which is simply due to the difficulty which the theorists experience in drawing distinctions otherwise than between phenomena. For in so far as the agent has qualities—is determined—instead of merely "carrying" qualities, it becomes itself one of the known differences; and as such, it becomes one of the elements for the uniting of which alone the agent is postulated.

The movement of the relativist doctrine is shortly as follows. There is the individual experience, which is an orderly and intelligible unity. Within this unity there come to be distinguished two sides, the side of unity and the side of multiplicity; or two elements working together, the subject and the object. Each of these parts is, by the common habit of men, regarded only as it is in the whole, so that the abstraction of one from the other is, in the same movement, their synthesis. But, for theory, the abstraction comes to be dogmatic; and we are given a subject which is itself only, and is not the object at all, and an object which is in no way the subject; and these two things are not one in any sense, but are for ever two utterly separate things standing over against one another. And these two strangers *somehow* conspire to make the individual experience. Therefore that *somehow* comes into question, because the result of their union is an intelligible whole—is, in short, experience. There must be a principle of the order—is it to be found in one of the elements, or in both; and if in one, then in which of them? And there must be a principle of the *suitability* of the one to the other—is it to be found in one of them, or perhaps in some third stranger? Here, it is evident, there come into question the whole articulation of the world we know, the rule of universal law, the unfailing possibility of finding causes for effects and effects for causes, the secret of the reasons for things. The very possibility of knowledge, the very ground and foundation of our world, is therefore to be sought either in the abstract subject or the abstract object, if these two elements are regarded as wholly strangers to one another.

Now the first instinct of everyone, as soon as he has accepted this uncompromising cleavage of experience into two self-sufficient factors, is to regard the self as a pure recipient, and to find the whole of the *reasonable* nature of things grounded in the articulation of the abstracted objects. "I am just mere I," says he, "and all the many things outside me, and even my dimmest and most intimate states, are the other things, and are not at all myself. For between me and them there is fixed the impassable gulf of *knowing;* and in knowing anything I rank it at once as the other thing. For myself, I am just the thing they all shine on; certainly the principle of the order is not in my blank surface, but in them, as and how I find them." But the first investigations which he makes into the avenues through which knowledge comes to a man are enough to disturb that simple belief, and to throw his speculation, by a natural reaction, to the absolutely contrary point of view; and he will find the purely relativist doctrine of knowledge to be much to his mind as the only thorough and self-consistent explanation which remains possible to him. This doctrine is grounded chiefly if not altogether on psychological considerations. What one perceives, it says, is not perceived as it is, for one does not know the object, but only one's own states which arise in presence of the object. Perception is partly false, in that the impression is in a different kind from the object, having proportions and degrees which are not objective proportions and degrees; and it is partly incomplete, in that there are only so many kinds of sensation, whereas the kinds of qualities may in reality be inexhaustibly numerous; and to the end it is only the ground for judgments which may be mistaken. Knowledge of things is not direct, but is mediate, and the mediation of a perverting and inadequate sense is an insurmountable barrier to the knowledge of what lies behind the sense impressions. Qualities may be known, but only a certain number of kinds of qualities, answering to the kinds of organs which subserve perception of them; and even though sense were not, as ours is, limited, still—because sense still would mediate—though endless qualities would be known, yet the substance of which they are qualities would still elude us. We know nothing otherwise than *relatively* to the limitations and mediations of which we learn in psychology. And this train of argument might be pursued as far and as ingeniously as you like. For all process must, for this point of view, deceive; and the process is not so simple as the theory uses to represent, nor is mediation at one single point alone; and as you cannot know the object because sense intervenes, so, because the sense impression is also object, even sensation cannot be rightly or adequately laid hold on, for every sensation is itself mediate and relative in various respects; or, in other words, sensation is not bare, but may be analysed. But the general principle of the doctrine is simple; and its most significant results are as follows.

We were to find the principle of the intelligible order of things—the secret of system in the world. Now it is plain that this principle and this secret cannot be found, by the doctrine which we are considering, in the world of outer things. The elements for knowledge which are given from the outer world are mere sequences of states of myself. Such sequences rise and fade with waves of attention. Their moments never exhibit a necessary articulation with one another, but only a factual coincidence, concomitance, or sequence. Their flow is utterly irrational and accidental. This quality exists with that quality, and we know *that* it is so, but not *why*. This event succeeds that event, and we call them effect and cause, but the unity of the two cannot be known, and we can only say that this companionship is just a way they have. If there be indeed a reason *in the object* for that concomitance and for that sequence, it is hidden in the recesses of the things as they are, and not as they burst upon us in their isolation; it is an unknowable reason and no reason. In the object, because each sensation of it is discrete and by itself, we can find no principle of reason. So, as de Vries says, the qualities of the individual are separate self-sufficient things, in every respect independent, and merely put together in certain combinations and permutations to form the image. And so, as Naegeli says, we cannot know organisms as they really are, nor make an adequate comparison of them, for we only know the heterogeneous qualities of them. There is no *necessity* in the conjunction of qualities, they merely exist externally side by side. And there is no knowledge of the principle of unity in those various parts which make up the individual, and this principle of unity must therefore be feigned as one of the parts externally controlling the others.

But the principle of the intelligible nature of things is found in the subject. A form of unity in difference is supplied to the formless material and utter difference of sense. That heterogeneity, without necessary articulation in itself of such a kind as we can know, is articulated by the forms of perception and of judgment, that is, by the fashioning activities of the self, in order to experience. In the abstract self, in the categories of its understanding, lies the secret of the world. Not of that world which is—but of *its* world; a partial glimpse, falsified utterly and forever by the mere presence of those avenues of sense and of those very categories by which the formless becomes articulate. For we have no reason to believe that the form of perception and understanding in the *knowing* stranger is at the same time the form of the *known* stranger—*that* is the centre of the theory; for the form of the *known* stranger is unknown and unknowable.

Certainly there is a saving clause, but the degree of its validity is, from the nature of the case, hidden from us. The subject is, to a certain unknown extent, and in a certain unknown manner, formally adapted to have its manner and its degree of knowledge of the things of reality. The forms of the two strangers have this one point in common— *Who made the world made me;* that is the last word of the argument; and its saving clause is its condemnation as a theory.

Now it need hardly be pointed out how foreign this product of metaphysical abstraction is to anything which can be called an evolutionary point of view. For, to an evolutionist, the terms of a relation are fundamentally one, one in origin, one in reason, being made for and by one another. To him their difference from one another is nothing but the diversity which is necessary to the existence of true unity, and which springs out of that unity. It is true that we find theories of biology which discuss the mechanism by which a *stranger* organism is fitted to a *stranger* environment. It is true that other theories give us elaborate mechanisms by which the thousand *stranger* qualities of the same organism are kept together and ordered, although they have nothing to do with one another. It is also true that we find theories of the acquirement of a *stranger* mind by a *stranger* organism. But in so far as these theories really regard the terms as strangers to one another—in so far as they make their abstractions dogmatic, they are not evolutionary theories, but are hypotheses which do not square with observation. For observation does not find one of the terms existing barely and alone or otherwise than as qualified by the other term, and one element in a natural system cannot even be described except as being the others.

In so far as the relativist doctrine affects theories as to natural relations—and it does so to a large extent—it denies the validity and completeness of the ordinary explanations of research. For it does not find in the qualities and successions of the object any intelligible, necessary or inner relations to one another, denying real system, or at least the knowledge of it. They are *not intelligible* because the *reason* of the relation is hidden in the thing of reality, or thing-in-itself, and because the qualities and moments of the successions are isolated appearances, bursting on us out of the dark. They are *accidental*, because, so long as we may not understand the relations, we can never say, "If this is so, then that cannot be otherwise;" for one need not multiply examples to show that events, qualities, human character, are only *necessary* and can only be depended on in proportion as they are intelligible and reasonable. They are only held together *externally*, in that the inner nature of the one is not at the same time the inner nature of the other. We are not, in our biological theory, to look upon the cause as the same as the effect, nor upon one organ as inwardly identical with another. The identity has been abstracted away into the dark region of the thing as it is, but as it is not for us; and mere external difference comprehends all that is given. And, given mere difference, there is not the slightest ground for reason to stand upon.

The abstract identity, or thing-in-itself, is just the anthropomorphic agent of hypothesis. It is always the result of dogmatic abstraction. Cleave the individual experience into two self-sufficient factors, and you must make two anthropomorphic agents, the unknowable *self* and the unknowable *thing-in-itself*. Cleave the individual organism into self-sufficient qualities, and you must make an unknowable "*vehicle*" for each of them. And because the reason for every part of a system is properly to be found in every other part, you must, when you dogmatically disrupt the system by your abstraction, hide the reason for every part in the arbitrary working of the agent. The system is what is to be explained. Deny its real existence as immanent in every one of its parts, and you must explain the appearance of system by the external compulsion of each part by one thing which is only characterised as not being one of the parts. Abstract from the inner identity of the planets, and you must give an angel to each to guide it in its courses. Define the qualities of the organism by the first postulate of biology, and the second must inevitably follow. The thing-in-itself is an alogical principle, in that it hides within the workings of its punctual and blank identity, the whole *reason* of the system over which you have appointed it.

This doctrine of abstract identity is extraordinarily unworkable when it is taken seriously. So long as you may postulate one thing-in-itself to every system of qualities, the matter is simple. But every system runs into every other— it is all in the end one system—and you can hardly stop short of the supposition that there is but one thing-in-itself. But, in becoming one, it has lost its significance to the theory. This numerical diversity of each noumenon is the most interesting feature of this hypostasis of the concrete universal as bare identity. It is demanded by the theory because there are many individuals, and you cannot have an individual which has not an identity of its own, a character of its own. But qualities come under successively higher individualities, and the same events enter into different patterns. On what principle were we to know where the x of $A\ B\ C$ ended and the y of $D\ E\ F$ began? Even out of school it is not easy to say what is one thing and what is many, and the least reflection connects without limit. But you cannot get over the difficulty by making the noumena into a coherent system. For if they are determined in this way and that, they lose their character as abstract identity. So that their abstract numerical diversity, which is their sole *raison d'être*, is an impossibility. That unity, punctual, blank and unconditioned, is derived from the self-identical unity of the abstract subject. For the latter, being the pure focus of its experience, and not being in any way that experience itself, being pure form uncontaminated with content, becomes the model for the noumenon. We have in it an example of a numerical unity apparently bereft of qualities, a featureless identity out of which differences and qualities do actually come. And this formal conception is transferred to the object, necessarily, as we shall see, because of the dogmatic abstraction of subject from object, so as to give rise to the anthropomorphic agent. Once make your abstraction dogmatic, and you cannot do otherwise, whenever you have to do with the explanation of system, than go to the agent, although the latter is in contradiction with itself and with all methods of knowledge.

For in every perception and judgment, and indeed in every sensation, the object reveals a twofold play of identity and difference. No two things are so much the same as to be indistinguishable in respect of *somewhat*, and that *somewhat*, even though it be only numerical, is a difference. Nor are any two things so different as that we should not distinguish them in respect of *somewhat;* and that criterion of distinction is the mark of the system within whose identity alone they can be compared. The universality of this logical distinction, its necessity to every judgment, and its reflection in the articulation of language, force it on everyone's attention. Therefore, in philosophy, as in biology, you must either recognise it as a twofold logical aspect of things, or you must represent it as a quasi-phenomenal difference.

If you choose the second course, your biology will provide an *idioplasm* that bears all the qualities, which, except in it, have no necessary or intelligible relation to one another; and the working of that agent will be, as has been demonstrated, for ever unknowable. Your philosophy, in like manner, will provide a *noumenon*, which is itself for ever hidden from us, but which radiates upon us its manifestations; and no one of these manifestations will carry us beyond itself. So, although the idioplasm is supposed to be quite phenomenal and the noumenon is supposed to be in no sense phenomenal, I class them together as *quasiphenomenal* pictures of a logical distinction; for they are both, for the purposes of their respective theories, unqualified, and yet are not ideal aspects of their respective systems. For they both pretend to be real; and as there is nothing real which is not phenomenal (which gives us the positive element in the *quasi*), so also there is nothing phenomenal which is not determined (which gives us the negative element). In such sciences as deal with *system*, you cannot evade the distinction between universal and particular; and the form of your science as critical, or as working with the hypothesis of the anthropomorphic agent, depends altogether on whether you regard the distinction as an instrument of method, or as a quasi-phenomenal difference. And I cannot think that the latter way of treating the matter is justifiable even in a science with a charter for its own sphere. It is to make use of an abstraction which obtains in no one of the ancillary sciences of biology. The form to morphology is nothing beside and beyond its parts and proportions. But philosophy will give you a self which is beyond its states, and biology an organism which is beyond and beside its qualities. And you are left with the insoluble problem as to how the abstracted universal is related—as if there could be a relation such as exists between particulars—to the abstracted particular. The whole method depends on a fiction which is not even a working hypothesis; it is a mere logical fallacy. If biology is to treat of *individuality*, we need a better form of doctrine than that of the agent. As Mr. Seth[1] has well said :—

"The particular as particular—the mere self-identical, unqualified particular—nowhere exists; it is the abstraction of a logic not wholly clear about its own procedure. And the thing-in-itself is simply the fallacy of the mere particular in another form. The mere particular and the mere universal are alike abstractions of the mind; what exists is *the individual*. All that is real is—not particular—but individual; and the individual is a particular that is also universal, or, from the other side, it is a universal—a set of universals—particularised. The two sides are always there, and each is only through the other. There is no existence which is not determined so-and-so, that is, there is no substance without qualities; and equally there are no qualities without a substance to which they are referred. It is the nature of reality so to be, and it is the nature of thought so to think. But the substance is not an existence—distinct from the qualities—something that can be separated from the qualities and known by itself. The substance exists as qualified, and we know it through its qualities. How else should we know it?"

Thus the fallacy of postulating phenomenal differences to answer to logical distinctions, the "thinking involved in corporeal conceptions," lies at the root of all our biology of hypothesis. The method is tempting because it is easy, it is dangerous because its results cannot be refuted, and it is useless because they cannot be proved. Biology, however, depends upon a working theory of individuality, and, except for the postulates which I have named, there can hardly be said to be any attempt at an expression for that conception in the science. In the study of the various biological expressions for the unity of the organism, and for its differences, we have one criterion of the first import-

[1] Seth—"Scottish Philosophy," 174.

ance by means of which we may test them. That test is, the harmony or the contradiction of those expressions with the ordinary methods of morphology and of physiology. In a word, our appeal is to fact. And, in a matter of this kind, we have not to do with numbers of facts; the heaping together of illustrations is useless. It is, however, important that we should abstract from no aspect of the facts with which we deal, and that we should invent none. For by the abstraction from real process, and by the invention of the hypothetical agents and of the hypothetical processes of adaptation by the transformation of species, the systems are made.

The latter portion of the hypotheses demands special notice, and special comparison with fact, because of the eagerness with which the doctrine of organic evolution is commonly received both as a statement of well established fact, and as an adequate form of explanation for the adaptedness and for the particular forms of animals and plants. We shall later see that its formulæ do not adequately serve the latter purpose; but it is a separate question, whether they are records of facts. Since the days of the doctrine of type, the history of biology has chiefly been the history of theories as to the transformation of species. According to those theories, the similarities of organisms are due to common descent, their differences from one another are due to the variously modifying effects of the more or less indirect influence of circumstances, and their adaptedness is due to the continuous action of environment on species, and of one species on another, through long periods of time. Starting from a common and obscure origin, the various races have progressively become modified and have progressively diverged from one another until those many and various forms of life which we know have in the end been formed. This progress, which, with some exceptions, is ever in the direction of greater complexity or organisation, greater specialisation of function and ever increasing adaptation to varying circumstances, is accounted for on various principles. A vague impulse to improvement is present in organisms. Or, the individuals of a race, on arriving in new circumstances, have experienced new needs, which have led to new actions; these, when they have become habitual, affect in a purposeful manner, by greater specialisation or by hypertrophy, the organs involved in them; and these changes in organisation have led to identical inborn changes in the offspring of the individuals concerned, so that the new needs, by modification of individuals and by the preservation of those modifications in inheritance, are progressively satisfied in the development of the race. Or, there is such an admixture and elimination of living substance from a number of ancestors, in sexual reproduction, as leads to a variation of the offspring, in certain respects and to certain degrees, from the image of the parent; a new organisation thus arises, and this new creature, if it is more suited in any respect to the given circumstances because of its variation, is able to leave behind it more numerous and better equipped offspring than other members of its race can do; so that the character of the whole race is changed in a certain direction, or so that a dichotomy takes place, some members following one direction and others another. I need not here do more than enumerate other "factors of organic evolution," such as sexual selection, isolation, and panmixia. They are all forms of events which are supposed to have occurred continuously or under certain conditions in the history of every race, and to have determined the transformation of the race into the form in which we now find it.

It is not easy to think of the transformation of species, apart from those several theories which have popularised the conception, nor to ask whether or how it has taken place, without coming to a conclusion in the matter by leading evidence which is worse than useless. It is usually implied that the transformation must have taken place, because by means of it alone can the various classifications and adaptations of organisms be explained. But the fact of transformation, even were every step of its progress before us, could not explain anything of all that. The fact, even when established, that a thing has vastly changed from its former state, is not corroborated by the fact that it has such and such adaptive qualities to-day; nor, in its turn, is the fact of the presence of those qualities illumined by the fact that the thing was not at one time what it now is. And it is certain that it is only this bare fact of change which is used by the theories of organic evolution. That this is so is shown by the presence of complete theories of the latter, together with almost complete ignorance of the course of the phylogeny of existing types. The facts of that secular development are a matter for the most unrestrained conjecture. The theories have to do with the explanation of adaptation, and of the nature of classification in general; they can hardly profess to be founded to any considerable extent upon the evidences for the transformation of species.

Whoever will take those evidences apart from the uses which the process is made to have for the hypothetical systems, will find not only that all that significance which has popularly been given to the process is gone, but also that what real evidence there is, is surprisingly small. One might fairly say that the evidence in support of the view that the highly organised species of our time have arisen from the lowly organised species of past ages is quite balanced by the evidence against it, and that one has in the end to rest on an overwhelming analogy in favour of the transformation. The theory is in a sense a *pis aller*. If the vertebrate did not come from some sort of invertebrate, then where did it come from? And human institutions, language, character, and other such systems make it impossible to believe that there is anything organic which does not come out of simple beginnings and proceed, in course of time, to a higher organisation. One might even adduce the analogy of the development of the individual organism, were it not that prevalent theories represent the body as more complex in the germ than in the adult image. I am utterly prejudiced in the matter, and cannot think of the origins of animals and plants except as taking place by the transformation of species. But apart from that prejudice, I should not be persuaded by what evidence there is before us.

For, so far as we know, the process does not take place at present. I suppose that artificial selection is more potent, and has more striking results, than any other factor in the change, and it must work with an infinitely greater rapidity. But, so far as I am aware, artificial selection has never established a species of its own, and has never established a variety which will not return to that from which it came, as soon as it is set free from the artificial selection and conditions. The answer to this objection is familiar. It is that the artificial selection has not had time, in the hundred or the thousand generations through which it has acted, *radically* to alter the organisation of the species. It is still *fundamentally* the same, and only *superficial* alterations have been made in it. But it is difficult to see what those distinctions mean. We can get the most extraordinary changes in domestic animals and plants. Pigeons, dogs, and roses offer extreme examples. If extent of difference in very various qualities were our criterion, then the greyhound and the spaniel seem considerably different. The difference is as great as any which one can imagine that any non-artificial factor could make, and there seems to be no reason for supposing it to be a difference of a special kind, such as would not be made by the natural factors which are not due to human intervention. In what sense then is the difference not radical, except in the sense that it obliterates itself, while the supposed differences made by natural factors do not so obliterate themselves? And there seems to be no meaning in the common argument that the mere length of time for which a quality has been possessed by a race can make it more irrevocable and more "fixed." If the quality is there, it is there; and if it is not, it is not. But what does it mean when you say it is fixed, firmly rooted, or fundamental? None of the qualities which we have known to be in some way acquired by artificial selection or by changed conditions are, in that sense, fixed, even though a hundred generations have included them. The species, when left to itself and to normal conditions, will drop them in a few generations. And it is not easy to see how there can be a difference between qualities which have been preserved for three hundred years in a species by artificial selection and qualities which have been preserved for n years by natural selection. Indeed, the distinction of *fixedness* would seem to exclude qualities which have been acquired, and superimposed on the organism, at any time, however distant. To say the least, nothing which we know of the behaviour of species at the present day supports the theory of transformation.

The evidence from classification and from embryology is not more convincing. It depends on the presence of homologies in structure which are not accompanied by analogous function. For such points of resemblance are not, it would appear, due to any resemblance in function or to any identity in respect of their significance to the individual; they are rather accidental to the actual needs of the organism, and are therefore, because no other reason for them can be found, to be ascribed to a common parentage. For this reason "vestigial" organs in the adult, and the homologies of the various stages of the embryo to the adult forms of more generalised types, are regarded as especially significant to the investigation of the relationships of species. Thus, such a structure as the vermiform appendix, which has apparently no special function in regard to digestion, is the "representative" of a former large and functional diverticulum, and proclaims a relationship with the other mammalia the more clearly that it has no function, but persists solely through a sort of inertia, so that no reason can be found for it but this bare reason that it persists from a time when it was of some service to the organism. Again, the branchial slits in the mammalian embryo do not in any way subserve respiration; they are not, we are to believe, necessary features of a necessary stage in the development of the mammalian form; there is not, in other words, any reason for them. Therefore, they are the mere *vestigia* of the past, persisting by no reason at all, and demanding for their explanation the theory that the lung-breathing mammal has arisen from the gill-breathing ancestor of old times.

But not rudimentary and functionless organs alone are the points of this evidence for transformation. All homologies, in so far as they are not necessary to the function of the part especially, point in the same direction. That homology does not and will not fit with analogy—I use the words purely in their morphological sense—has always been the stumbling stone in the way of biological speculation, whenever the latter attempts a theory of the *rationale* of form. Goethe, Geoffroy St. Hilaire, and Lamarck all discuss the matter, and it has often been studied since their day. As they stood helpless before it, so do we. It may be said to be a great part of the chief of our problems. It is the point at which the doctrine of design and every philosophical study of the teleological development of the individual breaks down. The structure of the organism is not to be accounted for as having been determined for its functions. Hartmann's unconscious, if it had made this organism, would not have gone out of its way to follow a type of structure which is not necessary to the functions which it has to fulfil; it would not have combined two patterns in its work, one—the morphological pattern, having a hundred points of resemblance to a type, although those points are not pertinent to the actual functions of the species, and the other—the pattern of structure as it is organic and necessary to the functions of the whole body. For these two patterns do not fit together.

Now the morphological homologies have often been compared to the mere whim of an artificer who has pleasure in following a certain type, making the best of it, as it were, for the present purpose, modifying every element and proportion in it, yet forsaking it wholly at no point. That whim must not be confused with the efficiency of the artificer in making every part suitable to every other part and to the destiny of the whole object of his art. His caprice may not interfere with his efficiency, but the latter does, at least, not require the former. A man—we are to suppose for the moment—would be quite as good and quite as possible an organism, although he had no vestigial parts in adult life, and passed through no vestigial phases in embryogeny. For these are of the morphological pattern and do not belong to the physiological pattern, in so far, at least, as we know or have conceived of the latter. Now the conception of the artificer as mingled efficiency and caprice has been loudly condemned, chiefly on this ground, that the caprice does not belong to the efficiency, and that the metaphor of the artificer had reference only to his efficiency.

It is, however, interesting to see that the doctrine of common descent has not eliminated this difficulty. It is not too much to say that what is inherited is inherited and worked out in the individual by processes which all affect one another. The morphological pattern arises out of the germ together with the physiological pattern, and it is no more possible to say that any point in the former is *merely* inherited than it is to regard functional modifications of the type as *merely* inherited by themselves. The endeavour of true science would be to see the two patterns in one and not to ascribe a different origin to each. If the adaptedness of parts to one another (that is, the efficiency of the artificer) be due to their functional relations in ontogeny, the morphological structure of the individual in its every point (that is, the caprice of the artificer) must be in functional relations with the rest. We cannot make the whale's muscles adapted, and make of its baby teeth *mere rudiments merely inherited*. They all grew up together from the germ, and if the teeth had no *necessary* place in the whole life, we may safely say that they would not be there, however many generations of toothed ancestors had swum about in the water catching insects. In other words, in so far as the development of the individual and inheritance are concerned, there is no such thing as a vestigial or functionless structure. And we cannot admit that there are such portions of an organism as *have no other reason for them*, and *therefore* must be ascribed to the inertia of mere inheritance. Are the gill slits of the embryo there *simply* out of old habit, *simply* because the ancestors millions of years ago had them? Or are they there because they are a necessary feature in one moment of the development of the mammalian form?

In a word, the mark of the artificer's capricious adherence to type, the functionless structure of mere inheritance, or the morphological pattern in so far as it seems unnecessary to the physiological pattern, is to be regarded as in truth as efficient, as functional, and as necessary as anything else in the organism. And when it is so regarded, I think that morphological homology ceases to yield immediate evidence of community of descent, or that kind of evidence which it is at present supposed to yield. It is conceivable that there may be reasons why different groups of animals should present homologies either in the adult or the embryonic image, and that those reasons may be such as to render unnecessary the theory of a common descent. But in any case, those reasons are prior to that reason of the mere inheritance from past times. One may well wonder at those who talk so fluently of the presence of a structure in the unimaginable past as an adequate reason for the presence of its very considerable "remains" in the descendant of our day—and that although, on their own showing, the race is continually presenting variations, and the structure in question is unnecessary to the other parts of the organism. How, in their organism, which is but a collection of qualities put together (*zusammengesetzt*), shall a useless organ persist through ages of which history has only seen a day? For the "rudiment" has run the gauntlet of variation in every generation, and of the "struggle of parts" in every ontogeny, together with the cessation of natural selection and use inheritance and I know not how many other fostering principles. Yet through these ages it has been jogging along somehow among the other parts of an organism to which it has no relation of necessity or function —how?

If, on the other hand, we find some meaning in the relation of parts as necessary to one another, we find at the same time some difficulty in conceiving that any and every kind of organism can exist. If we are content to regard the organic differences as externally put together, each having a separate external significance of its own, then no considerations of possibility will stand in the way of an unreserved acceptance of the doctrine of descent. It would be unprofitable to discuss the matter, for the argument would be the mere balancing of one set of conjectures against another. But I do not think that those whose sole method is conjecture have sufficiently considered that although the ancient and modern species which we know—those, namely, which are at the ends of the branches of the supposed tree of descent —*have proved their possibility*, yet those hypothetical forms which are in the course of the branches have not made that proof. It may be said that some at least of the ancient forms which we know are in the course of branches, and not at their ends—that, again, is merely a matter for conjecture —but such is not the case at any rate in the great majority of palæontological organisms. And if modern species are thought to have descended from known ancient forms, the possibility of the intermediate forms remains a question. " Modification " is a satisfactory answer only so long as you do not go into the steps of its progress, and the imperceptibility of its degrees does not really hide the fact that there is a great space between first and last—a space which has to be filled with possible and viable creatures, "chords of nature," as the phrase goes. Instances, which seem to me quite insuperable, are, the modification of branchial appendages of insects into wings, the gradual acquirement of wings by reptiles to become birds, the assumption of a land life by aquatic animals, and the acquirement of the faculty of "complete metamorphosis" by insects. On the whole, the evidence from classification and from embryology in favour of the transformation of species is not convincing, while considerations of possibility raise insuperable objections.

The evidence from palæontology is of a different nature. It reveals that species have succeeded one another on the face of the earth in the order of their organisation. And here there enters what is known as "the imperfection of the geological record," which has obliterated all traces of the vast majority of the supposed ancestral and intermediate forms. It would be impossible for anyone but a professed geologist to estimate the value which should be given to that imperfection, and so to judge how much one ought to infer from the absence from the rocks of great numbers of steps in the transformations. So much use is made of this imperfection of the record by the arguments in support of the transformation of species, that one becomes suspicious of it because of its very convenience. What is left of the record does, however, give the succession, though *not* the transformation, and since it is impossible to conceive of any other origin for the successive species, one must simply suppose that each arose, by modification, from those which preceded it, and that some of the latter have been lost and others preserved.

Two of the most significant recent tendencies of biological speculation should be mentioned. The great facility with which evolutionary theory explained organic forms was largely due to the unlimited allowance of time for each change, and to the belief that such changes took place "imperceptibly," "gradually," "by degrees." With advancing research the time which can be allowed has become more or less definitely limited, and some have conjectured that the allowance of time is insufficient for the various changes, unless the rate of change was formerly vastly superior to that of any such changes of which we have knowledge at the present day. I cannot judge of such conjectures, both because I do not know what the data for them are, and because the present rate of change appears to be *nil*. And, on the other hand, the highest authority on variation regards the changes as having taken place, not continuously and imperceptibly, but *per saltum*.[1] Again, the doctrine of descent is entirely based on the analogical similarities of species, which are represented as degrees of relationship. But not only is an agreement as to the actual relationships of forms in a great number of cases absent, but a polyphyletic origin has been ascribed, in more than one case, to the species in a group. Whenever that is the case we have to do with such a degree and such a kind of similarity among the species concerned as leads to their classification together; yet that similarity is independent of the supposed relationship in descent. These movements of biological speculation appear to be of the highest significance in relation to the doctrine of organic evolution, and to be such as to increase its difficulties, as well as to limit its efficiency in explanation.

When its spurious significance has been removed from the transformation of species, the evidence for that process is partly purely conjectural, partly contradictory, and wholly

[1] Mr. Bateson.

difficult to estimate. But the hypothesis stands firm on an overwhelming analogy with the origins of other systems; and whoever should deny it as a fact would find it difficult to say where species came from, if they did not come from one another, and, ultimately, from the slime " in some warm little pond."

CHAPTER II.

THE FIRST POSTULATE OF BIOLOGY.

THE first postulate of the systematic biology is, that the qualities of the individual are separate constituent elements of which the organism is the total sum. The qualities are thus independent of one another as regards their significance, maintenance, development in the individual, existence as latent, inheritance and variation, and acquirement by the race. This postulate at once demands and makes possible the theories of the anthropomorphic agents in the individual, and the doctrines of adaptation by the transformation of species.

The fact of its authority over the biological systems may be found in the open statement of the doctrine by the more critical theorists, and it may also be found as implied in the work of others, and especially in certain definite conceptions and arguments which arise out of it. And I shall seek it in these various ways, beginning with the most notable examples of its dogmatic assertion.

De Vries, in his work entitled "Intracellular Pangenesis," opens his argument, as is rarely done in biology, with a clear statement of method. He at once finds it necessary to make an exact definition of the nature of organic qualities and of their significance to the individual, and bases the whole matter upon a frank statement of our postulate. The following words, taken from his opening pages, could not be more clear upon the matter :—

"Among the many advantages which have given to the doctrine of descent so great a significance for the investi-

THE FIRST POSTULATE OF BIOLOGY. 47

gation of living nature, a foremost place must be given to the shaking of the old conception of the species. For formerly they treated every species as a unity, and the sum of the specific qualities as an image having unity (*einheitliches Bild*). But an examination of specific characters in the light of the doctrine of descent soon shows that they are put together out of single factors which are more or less independent of one another. Almost every one of the latter is found in many species; and their changing, grouping and combination with the rarer factors, conditions the extraordinary manifoldness of the organic world. Even the simplest comparison of the different organisms leads, under this point of view, to the conviction of the composite nature of the characters of species. Thus the power of producing chlorophyll, and of decomposing carbon dioxide in the light by means of it, is plainly to be considered a unity which gives its peculiar stamp to a great part of the vegetable kingdom, but is wanting in many groups distributed through the system, and is, therefore, combined with other factors of plant nature in no inseparable manner."

After other instances of the special formation of chemical substances, the author proceeds as follows :—

"In like manner we must also assume the possibility of the separation of the morphological specific characters. It is true that morphology has not yet nearly reached the point at which it could complete such an analysis for every individual case. But the same leaf form and the same coarser and finer serrations of the edge of the leaf repeat themselves in many species, and even our common terminology shows that the images of all leaf forms are put together out of a comparatively small number of simpler qualities. It would be superfluous to add many examples, for any one can easily find them for himself. It is only important to accustom oneself so thoroughly to this conception that one ever sees clearly into the compound formation of the image out of its individual parts. It is then plain that the character of every individual species is put together out of many hereditary qualities, by far the greater number of which reappear in countless other species. And although so great a number of such factors is necessary for the construction of a single species, that we almost draw back from the consequences of our analysis, yet it is clear on the other hand that the number of individual hereditary qualities which is sufficient for the construction of all organisms is small in comparison with the number of species. Every species thus appears to us as a very complex image, but the whole organic world appears as the result of countless different combinations and permutations of relatively few factors. These factors are the units (*Einheiten*) which the science of inheritance has to investigate. Just as physics and chemistry refer to molecules and atoms, so must the biological sciences press through to these unities, in order to explain the living world out of their combinations."

Further, the development of the race in history is to be ascribed to the addition of new unit qualities to those which have already been put together to form the species. "In every case we see how one and the same hereditary quality may come to be bound up with the most different other hereditary qualities, and how, by means of these exceptionally varied unions, the individual specific characters may come into being." And the same explanation is made for variation. "Almost every quality can vary independently of the others."

The qualities themselves do not, according to de Vries, form any such unity as was supposed before the "doctrine of descent shook the old conception of species." The specific form, that is, the individual, is no longer to be conceived as having even any morphological unity. The separate measurements of form do not exist in proportions such as were held to be the very meaning of form. The various qualities or differences are only "factors," which are externally held together so as to give the delusive appearance of a unity in difference. Everything which can be distinguished from the rest, and which may exist in individuals of different species, is itself a complete thing. It, and not the organism, is the true individual. It is the unit to which biological investigation must advance.

"Even our common terminology," says de Vries, "shows that the images of all leaf forms are put together out of a comparatively small number of simpler qualities." His meaning in this seems to be clear. To qualify an organism in such-and-such a way, is to distinguish one separate thing existing among other separate things. And a true classification, if we could only get it, would be a description of these "comparatively few factors," supplemented by a description of their distribution among organisms. Thus *redness* exists in these plants, *hairiness* in those, and *a height of six foot* in those others. And such separately existing and, to physiology, accidentally combined qualities, are the real individual unities, which we must, in order to biology, dissociate from the accidental colonies in which they live in a sort of symbiosis. In a word, the abstract universal is the real unity, and the concrete universal is only a puzzling complication of our problem. And classification should not proceed by the analogical comparison of organisms, but by the enumeration of these abstract universals.

A more subtle and insidious approach to the same necessary postulate for hypothetical systems of biology is given by Naegeli, in his great "Mechanisch-physiologische Theorie der Abstammungslehre." He is impelled to create, since he cannot find, a sphere in which all organic qualities should be units, separate from one another, and capable of enumeration. And, when he has created it, he regards it as the true organism, of which that which we know is only the baffling and delusive appearance. I quote from the section entitled "Idioplasm as vehicle of hereditary *potential qualities (Anlagen).*'

"The comparison of different organisms suffers not only from our want of knowledge, but also from the absence, due to dissimilarity of organisation, of a common measure which should assign the value and therewith the true difference (of qualities). Thus we are unable to compare the fungus, the fern, the pine, and the fruit tree otherwise than by pointing out that in one of these plants there is a quality *(Merkmal)* which does not occur in another, and that in one case it has this nature *(Beschaffenheit)* and in the other case that other nature. But the difference can in no case be expressed as quantity, and thus as plainly presentable magnitude. Therefore all systematic distinction and estimation is more or less arbitrary, and all inference from them, for the purposes of phylogenetic theory, is hypothetical.

"There is, however, a condition in which the qualities which cannot be exactly compared and estimated are eliminated, and in which all organisms are reduced to a similar form and structure. It is the first and unicellular stage of development, for in the ovum all animals and plants are similar. But the ovum contains all essential *(wesentlich)* qualities just as much as the developed organism, and organisms differ from one another not less as ova than they do in the developed condition. The species is as completely contained in the hen's egg as in the hen, and the hen's egg is as different from the frog's egg as is the hen from the frog. If it seems otherwise to us it is only because many distinguishing qualities may be seized in the hen and the frog, while the distinguishing qualities in the ovum are hidden from us.

"The states of the ova are the short beginnings of the individual histories of development, and as such may be compared, as it were, with short segments of different curves. The short segments appear to be all similar and to be all straight, although in them the essence and the mathematical formula of the different curves is as sharply expressed as when they have been lengthened so as to show their characteristic differences even to the naked eye. Therefore the ova should be the true objects of comparison, for they would give us all the differences in the same form, *and therefore capable of being given in terms of measurement.*"

In this short argument Naegeli formally reduces the qualitative differences of the organism to differences which can be measured, and are therefore units and separate from one another. For if they are in this hypothetical manner qualitatively similar to one another, they can only be distinguished from one another, as Naegeli goes on to distinguish them, as numerically different things. We have been conducted to the germ in order to the demonstration of this state of the qualities, but when that has been accomplished the author freely treats of an existence of this kind for them in the succeeding stages of development. Each quality and each part is referred, not to the others, but to its "vehicle." And the qualities, separate from one another and measurable, *in* the germ, are not the qualities *of* the germ, but are the qualities of germ and every succeeding stage as *Anlagen,* or as *actual* existences answering to *possible* qualities. It is not necessary to go further into these theories in order to show how, when they mention the matter at all, they immediately proceed under some powerful impulse to describe organic qualities as constituent units which are numerable, and which do not find their ground of explanation in one another.

Now, it is evident that when once such a point of view is given, observation under it will do nothing to correct or alter it. It is certainly necessary to such an author as de Vries to give some explanation of the obvious fact that many qualities vary together and are inherited together, but it is not difficult to invent a theory to explain this binding together of the separate unit factors into groups. For it belongs to the method of such theories to ignore any examples of what they call the correlation of qualities, except those which are only occasional and those which arrest attention because they are not understood. For "common terminology" is the guide as to what is one quality and what is more than one. And that criterion is untrustworthy, because, when it sees two qualities which are invariably combined, or which even have an intelligible relation to one another, it very rightly calls them *one*. In fact, we have not any serious attempt to obtain a true analysis of the organism, on the principle of de Vries, just because "common terminology" is the final arbiter.

And it is the same guide which leads our theorists to look on description as the analysis of the individual into what are practically subordinate individuals. A thorough attempt to complete the analysis and to count the constituent factors would at once clear away this postulate from the science. But, in fact, the postulate is really offered to us as an account of the methods of observation; it is a thinly disguised psychological account of the way in which our impressions are formed, and no theory of the nature of organisms. It is an attempt to justify "common terminology," and no attempt to penetrate behind it. And with such a theory as that, observation has certainly nothing to do.

But what qualities come under the definition of this postulate? Are they merely morphological, or merely synchronous qualities? We shall find that, in fact, every difference—everything which can be distinguished from the rest of the organism—is to be a separate constituent element. The stages of the development of the individual are to be discrete, and without intelligible relations to one another, no less than all the qualities which may exist together at one time. Morphological and physiological qualities are thus independent among themselves and of one another. Even latent qualities have a separate existence from the rest. Alternative qualities do, in mere fact, appear instead of one another, and do not appear together, but they are not on that account related to one another. And the parent is unrelated to its offspring, for they are both the mere puppets of the anthropomorphic agent. We have to do with a doctrine of organic differences, *as differences*, and without much respect to what differences they are. And this doctrine may be seen in many common biological conceptions.

Perhaps the postulate is associated with no theory so closely as it is with what is known as the cell theory, and with those conclusions which are drawn from it for biology. For those conclusions are of such a kind as is not warranted by the mere fact that the body is throughout made up of cells and of their products, and that its growth, its repairs, and its hyperplasias are due to the multiplication and differentiation of those elements. No one would deny the general truth of the statement that there is nothing living which is not one cell or a colony of cells, but one would be slow to admit that the statement is, for biology, at all an adequate or even an important one in relation to the expression of the determination of form. And I can only derive its great popularity with the hypothetical biology from the fact that it may be, and commonly is, so stated as to evade the problem of the form of the individual altogether. It is possible so to insist on the multitude, on the similarity, and on the independence of cells, as to deny the supreme individuality of the body. This whole organism, it is said, is but a colony of these, the true individuals, and the secret of its form is to be found in their habits of growth, reproduction, and differentiation. And so the question of the whole and the parts is removed from the sphere of the body, in which we have some opportunity of studying it, only to be repeated in the microscopic sphere of the individual cells. But, in the latter sphere, the question does not really come before us, so that the whole form of the argument does no more than to evade it altogether. For the cells, in their turn, are regarded as the mere stones out of which the living edifice is, externally, put together.

The point of view is almost universal. Perhaps Virchow's

vigorous advocacy has done most to strengthen it. He goes so far as to say, for instance, "If it is possible to separate certain elements or groups of elements from the association of the human body without their ceasing to show the features of life and to maintain themselves, it follows that this association is not, in the traditional sense, a unity, but is a company, or rather, a society." (Virch. Arch. f. path. Anat. u. Phys. IV. 378.) And again, "*Every animal shows itself to be a sum of vital unities*, every one of which manifests all the characteristics of life. . . . The structural composition of a body of considerable size, a so-called individual, always represents a kind of social arrangement of parts—an arrangement of a social kind, in which a number of individual beings are mutually dependent, but in such a way that each element has its own special action, and, even though it derive its stimulus to activity from other parts, yet alone effects the actual performance of its duties." (Virch. Cell. Path. 13.) Huxley, again, calls the body "an aggregation of quasi independent cells," and regards such a definition as a serious aid to the solution of the problem of form. And Darwin introduces his provisional theory of pangenesis by a well-known passage to the same effect. Plainly, the matter is considered as of great importance in this connection.

We have abundant evidence of the same point of view in the ever recurring metaphor taken from human society. For as the form of the latter is given is the cooperation of so many millions of separate and independent persons, who are to be the true social individuals, so is the form of the body to be given as the sum of so many millions of "quasi independent cells"; and thus the politics of the extreme radical are transferred to the organism. But the instruments of the constitution, the institutions of the country, the character of public opinion, the price of bread, and the pieties and observances of our common life are the features of society; and we are not our own, but are of their creation.

The "quasi independent" man goes to prison. As an abstract unit, a man has nothing of society in him at all. Even if cells were as independent of one another as I am of my neighbour, one might well refuse to give any value to the metaphor. Whatever else it may illustrate, it certainly does not illustrate the organism as a mere "sum of vital unities"; and if it is seriously considered, it tends rather to support the conception of the body as "in the traditional sense a unity" than the reverse.

The cell is an individual, and as such it must be considered, as Virchow puts it, as *alone* effecting the actual performance of its duties, on receipt of a *stimulus* from without. The changes which proceed in it and from it are thus to be considered as *activity* called out by *occasion*, rather than as *effect* resulting from *cause* which contains the direct reason for the effect. And in activity is, to quote Bunge, the "mystery of life"; or, in other words, activity is the mark of individuality. Now, if the individuality of the body is to be slurred over, and all its problems are to be answered by a formal reference to the cell and its activities, we have a right to expect that some architectural principle should be found in the cell itself. Each of these elements, since they are not put together by an artificer, and since in them alone there is to be found the secret of form, must have, in some way, the form of the whole in its own abstract and independent existence. But I find no such attempt to fill up the conception of cells as anthropomorphic agents. The stimulus for the cell comes from the rest of the body. Its food, its conditions of tension and stress, its temperature, its very form, the periods of its divisions, and its wanderings —these are all not its own but from the whole, as soon as we come to serious research. Elements removed from the body do *not* maintain themselves; they must be returned, before they die, to normal conditions. It is, in the end, with the *whole*, and not with the *cell*, that physiology and pathology are occupied; and one may give its due to cytology, and yet doubt whether its exquisite researches can throw light on the economies of the body. In a word, we are not likely to find, within an individual abstracted from a system in which it is only an element, as thoroughly passive as it is active, the principle of the architecture of the whole system. "Of this organisation itself as such— that is, of the mechanical apparatus it presents to us—the microscope tells us nothing whatever. The microscope only enables us to see a single cell, a single germinal particle in connection with more or less of its own formed material —a single coral, so to speak, and the polype that died into it: it tells us nothing whatever of the vast machine which these polypes have all unconsciously built up with their coral. The mighty and complex frame of man is, after all, despite its innumerable parts, a unity; all these parts but go towards that unity, are sublated into it. Now, what of all that does microscopic observation tell us? Why, simply nothing. Myriads of miserable Egyptians carried stones to the Pyramid; but no microscopic watching of any of these, stone and all, would ever explain the Pyramid itself—*its many to a one.*" (Dr. Stirling: "As regards Protoplasm," p. 75.) I suppose that the cells of the body are at least as intimately dependent on one another as the planets are bound together in their system, yet who would propose to call that system "an aggregation of quasi independent units?" Who, again, would seriously treat of society in that way? But, in fact, the metaphor from society lies so easily in its place in these arguments, only because society is conceived as no system, but as practically anarchical, as indeed it *is* from the standpoint of the abstract individual man. To these arguments, "society" means a crowd rather than a unity; it is a collective and not a singular term. And in so far as it is so used, it is not a pertinent analogy for the unity of the organism.

Individuality is shown in those internal proportions and adjustments of which individual cells are the mere materials;

and the more free and independent those elements appear to be, the less they have to do with the architecture of the body. And this is the meaning of Dr. Stirling's paradox, that "it is the so-called dead formed material that alone truly lives, and not the so-called living germinal matter that is assumed to die into it." While research into the fine structure and the functions of these cells gives us light on the formation of that fragment of the body in which they occur, we do not and cannot gain from it any information as to the principles of the building of the body.

An interesting example of the way in which the doctrine of the independence of parts is commonly regarded as of value for the explanation of the form of the individual, is given in Roux's ingenious theory contained in his "Struggle of Parts." In amplifying the conception which was given by Thiersch of the histogenetic equilibrium, Roux deals successively with the struggle of molecules, cells, tissues and organs, and, in a somewhat Darwinian manner, derives the form of the body from the variation of its elements, or metaplasia, and from their conflict with one another. In so far as the adaptations of individuals to the environment may be explained by such hypotheses, the adaptations of part to part within one organism are of course susceptible of the same form of explanation. And Roux must be credited with one good service; in that he pointed out that the structures of the individual do not arise as it were by the miracle of inheritance, but arise by processes of action and reaction on one another. But because the processes of his hypothesis were not those which are known to research, but were the external action upon one another of what can only be called separate individuals, the theory must be condemned for dealing only formally with hypothetical processes and hypothetical, self-determined, abstract agents. For, if the matter came to be one of detailed observation, it would be impossible to judge on what principle the conflicting individual parts should be limited from one another as separate individuals. One might choose any one of a hundred ways of dividing the body, and be equally right in any one of them. There must be some curious fascination about this conception of *struggle*, that it should be introduced into the explanation of the parts of that which is the most perfect and unique unity we know.

The independence of qualities is further implied in the ordinary biological conception of variation. For it is commonly supposed that one quality may vary without an accompanying variation through the whole body; and it is mainly upon that supposition that the theories of organic evolution depend. For their success in the explanation of organic form depends upon the implied unlimited possibility of this and that quality being combined for various purposes in one organism. Now, if a certain change in the image may take place without causing and being caused by other changes answering to it, and if one state of the organism has thus no reciprocal relation with its other states, then indeed we may at once give up the unity of the organism, as Virchow or de Vries would do, as an antiquated delusion. For then an animal or plant will appear to us as a collection of qualities, which, except in certain rare cases which are specially provided for and defined, neither imply nor need one another.

Certainly, to a superficial view, it would appear as though there were such a thing as uncorrelated variation. Flowers may have a constant form, and yet vary greatly in colour. Changes of form may even take place, in some species, upon one side of the body and not upon the other. Again, the males of a species may vary in certain respects and in certain directions, although those variations have apparently no corresponding or correlated representatives in the females. And many examples of a variation of one period of the development of the individual, apparently without influence upon other stages, will occur to the reader.

Indeed, in the greater number of cases, variations would appear to be of one difference alone, and to be without effect on the whole.

It does not, however, follow on that account that we can regard the differences as in reality thus independent of one another, in their quantitative and qualitative changes. If one flower has been held to imply and to make necessary the whole world, we may expect to find that the parts of the flower are at least not less dependent on one another than the flower is on the rest. And it remains that we should see whether the identity of the body is such that a changed part means a changed whole or no, and whether we are not merely puzzling ourselves with phrases when we speak of independently varying qualities.

It is useful to remember, in the first place, that variation is not, as our biologists make it, a process of its own and by itself, which may take place in this part or in that, and work in this way or in that. It is not a principle working on an otherwise static image, or a change in a body, which, except for "variation," does not change. We may give up at once the hope for a physiology of variation, as a special branch of physiology, and for a mechanism for variation, that should be discovered among other cytological elements. To theories which ascribe to inheritance such a process and mechanism apart from and beside the other processes and structures of life, variation is, of course, just a special process and a special provision of mechanism within that of inheritance. Now, such a process of variation or mechanism for variation both can and must have effect in one part or quality and not in another; and it would be astonishing if it worked in all the differences at once, still working so as to suit them to one another. For if it should thus adapt its various performances to one another, that mechanism, being itself a part among others, would be effecting, by its own abstract and uniformed activity, a result of the kind which we have come to ascribe only to the cooperation of all the parts. And of those parts no single one can, in this unique respect, receive a pre-eminence.

But the body, in which we remark a variation, has grown up from its beginnings. It has not remained the same body, changed only in this respect or in that by an organ for variation and its functions. It has come, through changes in which every trace of the typical image has been lost, to a stage at which it wholly resembles the parent or type, except in respect of certain points. These nine points are common to it and to the parent, but that tenth point is not common to both, but is a variation. We cannot question that some of the points are quite in common, and that others, abruptly and, as it were, independently from the rest, are not at all in common. But the question comes to be of what nature those "points" are, whether the varied points are such as that we should get in them the *whole* difference between the one individual and the other, and therefore, whether we are warranted in saying that this and that quality have varied in complete independence of the rest.

The individual is complete and perfect, and there is no variation in it from some supposed form within it which has been expressly departed from or altered. Its variations are simply points of difference in the analogical comparison of it and the type. That comparison selects its points of similarity and of unlikeness, and does not by any means take in all the points which would be included in the exhaustive anatomical, histological, and physiological examination of the single creature. It selects them, further, on a principle; it has a criterion and a definite method for the comparison of organisms; and from that method it cannot escape. It does not take its points at random, but chooses them as having a certain significance to the organism; and it is in relation to that significance that they are compared. *Organs* are compared, not heterogeneous masses taken at random from the body. Morphological comparison is made in respect of definite structural proportions which have a recognised place in the whole image; they are compared as proportions, and not as single measurements. The comparison of colour on the wings of insects has for its "points" definite elements in a pattern. In speaking of inheritance and of variation, you name this part, this organ, this bone, this tissue, this spot of colour, and all of these points for comparison are chosen because each of them has a definite significance to the morphological or the functional unity of the organism. You are always comparing, as it has well been said, *in reference to a standard*, and on a principle. It follows, therefore, that you do not succeed, however many points you raise, in comparing the organisms with which you have to do, *in all respects*, just because you are always comparing them in respect of somewhat. There are differences and similarities between your specimens of such a kind that you would call them irrelevant to biological comparison, and, indeed, it never occurs to you to bring them up.

Since the comparisons which have been made the basis of the biological doctrines of variation are mainly, if not altogether, morphological, the fact that two organisms may differ in one point of this kind without differing in others of the same kind, has become enlarged into the statement that two organisms may differ in one point without differing in any others whatever. And, though this is not the most important fallacy in the supposition that qualities vary independently of one another, we must not forget that a variation even in colour probably means a profound functional variation throughout the whole body. I do not mean to say that it will ever be possible to tell what alterations in different organs, and what changes in the chemistry of the blood, and what thousand other variations go to the making of a change in a spot on an insect's wing, but I certainly think that such an intimate nexus between "points" in different kinds is supported by observation, and must be at least formally allowed for in theory.

The matter of colour alone would furnish innumerable examples of variations which, although apparently independent and unaccompanied by other changes, are yet due to a profound change in the whole organism. It would not be easy to give such examples from what is ordinarily called variation, because that term includes only new associations of qualities which do not change in the individual in question, and cannot be made the subject of experiment; the variation, in other words, is not in the individual, so that it can be observed, but is between two individuals, so that it eludes us. But the association of colour changes with other physiological changes is well seen in the history of many individuals, and the connection can hardly be of a different nature in one case from what it is in the other. There are remarkable colour changes which are associated with the maturation of the reproductive glands, and which may even come and go with the seasonal reproductive changes. The colour difference which is associated with the other sex differences in many species, and which disappears with the removal of the reproductive organs, is an instance of a quality which does not stand by itself, but which is intimately linked to differences in respect of innumerable other points; and all these differences are one difference, and the difference is the difference of the whole. Instinct, temperament, sensibility, blood, skeleton, brain, size, in these respects and in many others do sexes definitely differ; yet it is one difference, the difference of the whole. They differ, not merely in respect of this and that, but altogether; and however many and various are the points of difference which we take, we are still not at the end of it. Again, alterations in the quality of food will alter the colour of feathers, and a chemical difference of the blood is seen to belong to a colour difference in the skin. Morphological characteristics among men are associated with particular characteristics of temperament, and nutritional changes are even more marked in respect of such effects. The various faces of men tell us of much more than merely of the developments of this or that muscle. In the end, we have to do with differences throughout the whole organism. And if we knew insects as we know men and women, we should find every change in colour reflected through the whole creature. It is just a question of how many and how concrete and significant are the points for comparison to which we have access. In the case of the unity of human character, of which we know most, the unity of all characteristics is never questioned. One of a man's actions is often enough to give you his whole self, and the difference of one man from another is not merely a difference in respect of this or that isolated particular, but is a difference of the whole unity, revealed in this or that point according to the standard by which they are judged.

It is not otherwise in the case of organic variation. If the colours of an animal may be changed by removal of organs and by alteration of the composition of its blood, we may be sure that, with a certain "variation" in the organs, there goes a certain variation in colour. We may be sure that the individual is a unity in respect of variation as in other respects, and that when it differs from the type in this or that point it differs from it throughout, with a difference of which the single point is, for some reason, the only mark which comes to our knowledge. In a sense, therefore, there is no such thing as uncorrelated variation; I mean in the sense in which that term is in contradiction with the possibility of physiology; for so it is used by the biologists, who expand it into a doctrine of variation in general.

What then is the independence of differences as regards variation? For we have found that there may be only one point of variation in one kind, as, for instance, that just this spot of colour may be changed without any other changes, or that just these arms may be longer without longer legs or longer spine. And on the other hand numerous examples show that such isolated points of variation carry with them the most profound changes in other kinds. But the theory of the independence of qualities in variation has been largely based, not on the discovery that there was one point of change while *all* other points of *every* kind remained similar, but on the mere observation that there are points of difference, and that, in spite of this variation, there are still points of similarity. As, for instance, " though the arms are abnormally long, yet the legs are normal in length," and not, " though the arms are long, yet there is no other point of departure from the type." Or, " different colour with similar structure," and not, " different colour with everything else similar."

Now the independence of the varying point is just the abstractness of the kind of point in question. We compare, as was said above, in respect of somewhat—morphological pattern, colour pattern, particular function, what you will— but always *somewhat*. And the variation, or difference from a certain type, of the creature in question, being necessarily a difference of the whole creature, yet shows itself, in reference to any particular standard of difference, in all, in a few, or in no points under that standard. It is thus often present in a change in one part of the particular abstract pattern but not in other parts of that pattern, because other points of difference would be irrelevant to it, in that the other parts belong to it as they are. It does not seem necessary that they should all change in order to belong to it, and the fact of their not changing does not seem to show that they do not belong to it or that they are independent of it. The point of difference, far from being an external change of one part without relation to the rest, is the expression of the whole difference under one particular criterion. Thus what is characteristic of an individual,

THE FIRST POSTULATE OF BIOLOGY. 65

in ordinary biological comparison, is by no means the whole of it, and its characteristic difference is not its sole variation. Only, the rest is, for purposes of comparison, irrelevant. But it is not irrelevant to physiology; and it must not be dogmatically asserted not to exist. For biology pretends to rise above the mere abstract plane of comparison, and to give a theory of the nature of organic qualities and of classification. And, further, the apparently irrelevant tends, in the progress of research, to show itself relevant, and therefore to become characteristic. To a child, one or two leading points in a particular man, profession, or institution, are characteristic, and all the rest is irrelevant. As he grows older, an increasing amount of these contents becomes intelligible to him as belonging necessarily to the unity of each. Similarly, two flowers may appear to us as differing only in number of petals or only in colour. It is incredible that they should not also differ throughout, and that the latter differences should not necessarily belong to the first primary difference. Why, then, do these other differences evade us? Because, in so far as they are in the same kind and under the same criterion as the primary difference, they are included in the unanalysed content of that first difference, having an intelligible relation to it, as, to have big muscles is to have prominent marks for their points of insertion, or, to have long bones is to have long muscles; and because, in so far as they come under a different criterion from that of the primary difference, they must also necessarily escape attention, not entering, in any special significance, under that first criterion. If, for instance, the blood of a short man were compared with that of a tall one, or if the morphological proportions of an albino were compared with those of a dark man, I suppose that in either case we should be disappointed when we tried to make out a definite difference, necessarily attached to the primary difference of comparison. This would be owing to the fact that we should

E

be, in such cases, laying two patterns side by side, the chemical and that of size in the one case, and the morphological and that of colour in the other, and half expecting that a difference between the organisms which shows itself as a tangible and significant characteristic in the one scheme will show itself also as a significant characteristic in the other. It is altogether unlikely that it should do so. There will be a difference in that other scheme, but it will not be a characteristic difference; it will not have degrees answering to those of the first; and it might well be such as could not be dissociated, as the definite result of the primary difference, from all the other conditions of the peculiar chemistry of the blood in the one case, or the peculiar morphological proportions in the other. To a characteristic difference under one scheme, other differences, though necessary to the former, do not correspond as correlative characteristics under another scheme. And the characteristics, under this second scheme, though they are irrelevant to the first, are yet not unrelated to it.

The principle of uncoordinated variation is, in fact, assumed as possible, merely because its adherents do not often take it seriously and examine its consequences. Any variation which they offer in illustration of it always involves, under the same term, a host of necessary correlations which they do not distinguish from it. Thus a variation in the length of the legs is two or four coordinate variations in two or four legs. If the femur is long, so are the sartorius and the rectus femoris, and the sciatic nerve, and if it is short, they are correlatively short with it; yet this is all called a single uncorrelated variation in the length of the thigh. We do not debate about the origin of each of these variations separately, because the relations of necessity between them are evident, and there can be no talk of independent variation except when those relations are *first* obscure and *then* denied. When they are obscure it would be saner to seek for them than to calculate the chances of their all occurring together. If we are to ask how the sartorius comes to vary with the femur, we may also ask how it comes about that the skin fits the body, or that the colour fits the skin. These correlations of variation are intelligible, and therefore one part is not said to vary alone, nor, when the correlation is evident, is an obscure biological principle called in to secure the harmony. But when the relations of a part are not fully known, our ignorance is elevated into a doctrine of the independent variation of parts, and when changes occur in unexplained conjunction we see in it the exceptional working of a mysterious principle.

Darwin's instances of correlated variation are almost entirely cases in which the nexus between the "points" is unknown, or, as he says, "mysterious." The principle is exceptional in its working; it is not recognised as inevitable and as intelligible. And the ordinary examples of it are not considered at all. He says, for instance, "Important changes in the embryo or larva will probably entail changes in the mature animal. In monstrosities the correlations between quite distinct parts are very curious; and many instances are given in Isidore, Geoffroy St. Hilaire's great work on this subject. Breeders believe that bony limbs are almost always accompanied by an elongated head. Some instances of correlation are quite whimsical; thus cats which are entirely white and have blue eyes are generally deaf; but it has been lately stated by Mr. Tait that this is confined to the males. Colour and constitutional peculiarities go together, of which many remarkable cases could be given amongst animals and plants. From facts collected by Heusinger, it appears that white sheep and pigs are injured by certain plants, whilst dark-coloured individuals escape. . . . Hairless dogs have imperfect teeth; long-haired and coarse-haired animals are apt to have, as is asserted, long or many horns; pigeons with feathered feet have skin between their outer toes; pigeons with short beaks have small feet, and those with long beaks large feet. Hence if a man goes on selecting, and thus augmenting, any peculiarity, he will almost certainly modify unintentionally other parts of the structure, owing to the mysterious laws of correlation." ("Origin of Species," p. 8.)

This is a museum of the most "curious" or "whimsical" or "mysterious" cases one could find; the relations of the parts of monsters, the relations between colour and deafness, and between colour and immunity from poisons being unknown, are given a special place as instances of a special principle. I do not mention the others, because they have a *prima facie* appearance of having an intimate reciprocal relationship. But they are gathered together as correlated variations for one reason only, that the *nexus* is not understood. Research has not proceeded so far. Therefore we are to have a *nexus* of a unique kind, and one, it should be remarked, which does not obtain between other variations. For the latter are not analysed into their parts, so that they appear to be *units;* and they are universally understood to be uncorrelated with others, so that they appear to be *independent* units.

The difficulties with which the principle of natural selection has met in this matter have been great. Each modification of one part is useless, even deleterious, unless other parts vary also in such ways as are necessary in order that they should support the first. How are we to conceive that this complementary variation, in the absence of coordinate variation, takes place? Usually thus. Out of the whole new pattern there arises, by chance, one element. This remains in the species because it confers some special benefit. Then, by chance, another element, which yet is necessary to the first, sooner or later arises. This also is retained, and so the matter proceeds until the whole pattern is completed. The whole adjustment may take place by imperceptible stages. But when it is seen that elements which are necessary to one another must arise at the same time, and when it is seen that the number of necessary changes—if we must take them as independent and therefore as number—is very great, the odds against their all occurring simultaneously and in the right proportions are practically infinite. "It is no reply to this," says Murphy, "to say, what no doubt is abstractly true, that whatever is possible becomes probable, if only time enough is allowed. There are improbabilities so great that the common sense of mankind treats them as impossibilities. It is not, for instance, in the strictest sense of the word, impossible that a poem and a mathematical proposition should be obtained by the process of shaking letters out of a box; but it is improbable to a degree that cannot be distinguished from impossibility; and the improbability of obtaining an improvement in an organ, by means of several simultaneous variations all occurring together, is an improbability of the same kind." Yet by just this impossibility, repeated in every generation, of the fortuitous simultaneous rise of variations which are necessary to one another has the whole world of living creatures arisen, according to that biological hypothesis which has the greatest influence. I wonder that, with their accustomed liberty of thought, the theorists of natural selection have not announced that, by a chance variation, the faculty of coordinate variation came long ago into the world, and, for its manifest advantages, was retained by all succeeding species by means of a special granule in the nucleus.

This difficulty is emphasised in controversy by the adherents of the principle of use inheritance; but their theory does not wholly avoid it. They conceive that the changes which an organism experiences in its purposive reaction to changing circumstance are to a certain extent represented in its offspring by changes or variations of the same kind. Therefore, because any one change in the individual is not worked out alone, but is, in common experience, functionally supported by changes which are necessary to it, the variation in the offspring is not of one part abstractly, but is of all the parts also which work together with it. Variation is, under this theory, no sudden leap in the dark: it represents merely the changes which the parent has experienced. One might almost say that variation does not exist for it, but that it has to do with a simple inheritance in every particular—a mere growing on of the whole parent. There is not in every generation a new creation of a fortuitously dissimilar creature. The element of chance being absent, and all changes being functionally induced, the theory of use inheritance does not err to the same extent as that of natural selection, against considerations of necessity. Thus Mr. Spencer, in discussing an hypothetical enlargement of the horns of the Irish elk, enumerates some of the chief changes which such a variation implies. "That the horns may become better weapons, the whole apparatus which moves them must be so strengthened as to impress more force on them, and to bear the more violent reactions of the blows given. The bones of the skull on which the horns are seated must be thickened, otherwise they will break. To render the thickening of these bones advantageous, the vertebræ of the neck must be further developed; and without the ligaments that hold together these vertebræ, and the muscles which move them, are also enlarged, nothing will be gained. Such modifications of the neck will be useless, or rather will be detrimental, if its fulcrum be not made capable of resisting intenser strains; the upper dorsal vertebræ and their spines must be strengthened, that they may withstand the more violent contractions of the neck muscles; and like changes must be made on the scapular arch. Still more must there be required a simultaneous development of the bones and muscles of the fore-legs, since each of these extra growths in the horns, in the skull, in the neck, in the shoulders, adds to the burden which the fore-legs have to bear; unless the deer with its heavier horns, head, neck, and shoulders, had stronger forelegs, it would not only suffer from loss of speed, but would even fail in fight. Hence, to make larger horns of use, additional sizes must be acquired by numerous bones, muscles, and ligaments, as well as by the blood-vessels and nerves on which their actions depend." ("Principles of Biology," vol. i., p. 452.)

Mr. Spencer gives only a rough indication of the necessary coordinate changes, and no one could complete the account; but such an instance as the above is sufficient to put the independence of variations wholly out of the question. No one quality or change exists without those others, inexhaustible in number and complexity, which are necessary to it, and where a physiology is conceived as possible at all, there are relations of necessity. But, inasmuch as the theory of use inheritance also commonly conceives variation of one quality as *primarily* independent, and as only later and in the life of the individual working out its necessary complements in the rest, in order that what were primarily unrelated features of the organism may be handed on to the next generation as still independent and yet as having an external harmony, I cannot find that Mr. Spencer's favourite "factor of organic evolution" avoids at all more successfully than does natural selection, the confusions which inevitably await theory which moves by the disintegration of the individual into self-sufficient and primarily unrelated parts. In whatever way the theories of the third postulate make use of the first postulate, they are unable to reconstruct the unity of the organism. And we shall later find that use inheritance is, in the end, as wholly founded on the doctrine with which we have here to do as is natural selection.

This first postulate is further shown in the ordinary biological treatment of *functionless* parts, which are supposed to exist in their own right and in virtue of a separate inheritance; indeed, it underlies that whole doctrine which is the mere dogmatic assertion as of a fact, of the abstraction of structure from function. Consider, for instance, a merely vestigial and functionless element in the morphological plan of an organism, such as is often given to us in the fibula or in certain phalanges. It represents, let us say, a structure, which, in other forms of a similar type, and certainly in the ancestors of the form in question, is and was functional, but which, modified as we see it, is useless. It persists, therefore, by the mere force of inheritance; it shows the particular relationships of its descent, and its inheritance and long persistence is independent of that of the functional parts. For the bone is not necessary to the rest of the body, and it is in no sense one with it. If it were thus necessary to, and one with, the other parts, it would, although mediately, subserve their functions. If it had arisen and if it were maintained by any reciprocal reactions with them, and if it were therefore due to anything else than an immediate action of a mysterious process of inheritance, it would be at once the object and the means of their functions, and thus it would not itself be functionless. We might in this way start by saying that the structure is *functionless*, and pass on to regard it as *useless*, and finally deny to it any relations of necessity with the rest of the organism, implying that there may be in an organism useless and, therefore, independent parts; and though we should be doing no more than simply to articulate a movement of thought which is everywhere present in biological hypotheses, we should be utterly confounding two distinct conceptions. A functionless part of an organism is not useless, it is merely useless in a certain manner. It subserves ends, but it does not subserve this or that end. Nothing organic is functionless except for a special abstract point of view. The phalanx which does not project from the body, and the fibula which does not reach from end to end of the leg, nor enters into functional articulation,— these are not functionless except under those names, and except when each of them is considered as a thing by itself.

For abstract morphology they are functionless, vestigial, what you will, but they are not functionless in the way in which the doctrines of inheritance and of ontogeny take them to be so. We are told that these parts, being unnecessary, having no other reason for them, and being therefore such as cannot be maintained by the factors of organic evolution, are mere rudiments representing formerly functional and therefore formerly necessary parts; that they are therefore *merely* inherited; and that, being *merely* inherited, they persist in independence of the rest, and by no reason whatever. In such a view I can only find another example of the bondage in which the biology of hypothesis stands to abstract morphology, although it pretends, by inventing processes of its own, to take account of the processes which are known to physiology. For the parts in question are not foreign bodies. They are rather, in the sense in which the doctrines of inheritance and of the development of the individual use the term, functional, although not as phalanx and as fibula. They are built by armies of moving osteoblasts and by the transformation of cartilage, and they are fashioned and maintained by cells which destroy and by cells which deposit bone. Their membranes are supplied with vessels and nerves. They take up space, they alter blood, and they subserve feeling; they are not without their origins and their results. There is adequate reason for them in the rest of the body, otherwise they would be encapsuled with fibres or eaten up by bone-eating cells. They exist, and therefore they are necessary, and the body recreates them every day. Yet we are told that they are functionless. Yes, by analogy; but not as regards the economies of this body of which they are parts. And what has inheritance or the development of the individual to do with that analogy? It is what is there and what occurs that comes before us for explanation, and not, in the first place, the homology of what is there and occurs with something which is not there and does not occur. Is this structure a modification, a vestige, a remnant of this or that in another species? Surely not. *This*, that I point out with my finger, is itself, belonging to its own body, and having its own relations with the rest; and because of that belonging and because of those relations it exists. My hand is my own, it is no modification of yours.

When a part appears to be useless, it does so merely because we are considering it as one separate whole thing. Now, we only give it that unity (usually by analogy with a homologous structure) in respect of a certain use which it, as one whole thing, is supposed to represent. And when it is found not to have that definite use, we say without qualification that it has no use. Certainly *it* has no use, but we have not to do with it, we have to do with a collection of things, with mere dust to be formed into unities of function under another scheme than that which we have unsuccessfully tried. The separate wholeness which we attributed to it does not exist in nature, and we can hardly be surprised if a part of the body, arbitrarily marked off from the rest, does not appear to have any special reason for it as a separate whole. That morphological elements should be functionless is therefore not surprising, and is no ground for any further inference than that the morphological category is too abstract to serve as the sole category for the enquiries of biology. Nothing in the body is one structure except under an abstract scheme within which it has a function, and one may sub-divide and cross-divide the body into organs under very various schemes. There are, for instance, the ordinary morphological scheme, the histological, the cytological, and that of the germinal layers. And what is functionless under one scheme is not so under another. Certainly the vermiform appendix is "vestigial," but you would not deny function to its peritoneum, connective tissue and mucous membrane, nor again to the nuclei and centrosomes of its cells, nor would you say, on its account, that the hypoblast is partly functionless.

The conception of function is thus largely relative. It rarely comes near to exhausting or even to formally allowing for all the relations of a part, just because to regard the latter as one part delimited from the rest is to delimit it in respect of one function. Every organ, every part of the body, every histological system, and every kind of cell, must, to take the chief example, have its specific reaction on the blood, and, very probably, its specific contribution to the tone of feeling. The nervous system does not *only* subserve sensory motor reactions and its directly trophic influences, nor does muscle *only* subserve the relative movements of parts, nor is bone necessary *only* because of its firmness and resistance to stress. And the differentiation of cells does not prevent them from fulfilling all the general functions of cell life to a greater or lesser extent, except perhaps conjugation; at the least they all divide, grow, assimilate, and excrete. Every part has weight and takes up space. There is an infinity of relations; and every one of these, under some scheme, is a function, but is not considered at all under other abstract points of view. And some of these relations are disregarded, and others are distributed under the sciences of special research, on principles of convenience for hypothesis. It would merely create confusion to no purpose if cytology were to take into account morphological difference, or if morphology occupied itself with histological functions. But when we come into a sphere in which it is possible to predicate *absolute* uselessness of a part, and to go on to infer that it has no necessary relations to other parts, and that it is therefore merely inherited by itself because of a use which something like it had very long ago, it becomes us to take into account the manifoldness of relations. We must see whether there is any concrete principle on which we may call any part *one*, and, therefore, *as absolutely one*, functionless; or whether, in so doing, we are fallaciously making out of a judgment, which may be quite true on its own abstract plane, the ground for a general doctrine of the unrelatedness of parts. And it seems to be evident that there is nothing in the organism which is functionless or unrelated to the rest, though there are *features* of it which are so functionless and unrelated. I consider that biology, in treating of functionless parts, has treated of them as absolutely functionless, and that this is in contradiction with the possibility of research into real process; and the origin of the whole matter seems to be that the distinctions of abstract morphology are commonly held to answer to divisions between parts of the real organism, and that these parts are held to exist in their own right and independently of one another, as regards inheritance, the development of the individual, and the maintenance of the image.

Function is not a simple term. When it is used in general biological theory it is supposed to mean all the activities of a part, but in most cases it merely means that activity in respect of which the part is called one thing, and it is, in consequence, usually a purely morphological conception. In the science of physiology it has sometimes this meaning, and is at other times more concrete. From this indefiniteness of meaning there arises a difficulty with regard to the purposiveness of parts. For not all that proceeds from a part, but only one single external significance is held to be its end and its significance to the individual. That single and special function may be only exercised by the part once in its history, or at rare intervals, or it may even be a matter of chance whether it is ever exercised at all. Horns may never be used in fight, and pollen may be wasted on the wind. Such cases as these are only specially remarkable ones, for in the case of every structure it is fundamentally the same. And one is apt to assume hastily that the significance of the part to the individual has nothing to do with its rise and its maintenance in the individual, and this assumption, when it is generalised, becomes the law that structure preceeds function in the individual development. That whole movement of thought is due to the attribution of a merely abstract and external significance to the part. This rare or chance function, when it is not exercised, does not, for all the purposes of the study of ontogeny, exist; and in such cases as I mentioned above, even the fulfilment of the main function can have no reflex effect on the part in question. The part is, however, necessary to the body because of its inner functions; it is developed and maintained by its physiological relations with the rest. Just as a man is more than his external professional significance under the abstract scheme of professional life, and has wide relations with the rest of his community otherwise than through his office, so does an organic part arise from and subserve other and in this respect more important functions than those by which it is commonly named. And just as no one would think so abstractly as to condemn a man as functionless, useless, and without a *raison d'être*, merely because he did not fit into that professional scheme, so we shall err if we regard anything which is in the body, and has therefore intimate inner significances and relations, as functionless and as without *rationale*, merely because it does not effect a function which we ascribe to it by analogy under an abstractly morphological pattern. A *vestigial structure* is a conception for morphology, but for the purposes of a concrete doctrine of inheritance and ontogeny we have to rise above it, giving it a place indeed, but not thinking of any part as really, or as regards the individual, in any sense vestigial.

The postulate of the independence of parts is further to be found in the biological treatment of *latent* and of *alternative* parts and qualities. For we have had occasion to see that the latent quality has an important place in biological speculation; and a great part of the theories exists only in order to account for the supposed independent existence of alternative and other latent qualities. *Anlage* is a word which occurs on every page of the German authors, and it is to be remembered that the *Anlage* is not a thing which has ever been seen, but is that hypothetical object which represents the latent existence of one future particular. It may not always work out its quality, and make it appear, but so long as this or that feature of the organism is possible, so long there is an *Anlage* for it in hiding in the protoplasm; and such a *foundation* for a unit quality may persist through many generations waiting its opportunity for activity. The manifold features of the organism are *latent* in the germ, and indeed often in the somatic cell; and the regeneration of lost parts is due to the existence of the necessary parts and other differences in a *latent* condition in the embryonic tissue of healing. All organic differences are inherited in latency; out of latency they come into the development of the individual; they are represented as varying when latent; all the stages of ontogeny which have passed are latent in the adult; and each individual in an alternation of generations has the other forms of the cycle in a latent condition (*Anlagezustand*) within it. Each sex holds, as latent, the alternative characters to its own, and in more complex polymorphisms there is manifested in the individual but one set out of several sets of latent qualities. Every change which the species undergoes in new conditions was latent in it before. Certainly this matter of latency is the chief category of biology. It simply answers to all other differences of the individual or species than those which are apparent in the part or the stage or the individual in question.

Now, whatever is latent is simply not there, it has no existence. This plant will have flowers some day, or it will not—at any rate it *may* have them; but at present it is without them, and its flowers do not exist. But they are *its* flowers, they belong to its scheme; in a word, they are possible. Latency is possibility, and a thing is possible because of something else. And the problem for biology is to find a form for that *something else*. All the theories

THE FIRST POSTULATE OF BIOLOGY. 79

which postulate "vehicles of qualities" exist for no other purpose than to find the ground of the possibility of latent qualities. Now, the latent is either possible because of the whole of the organism of present appearance, that is, it is implied in the differences which at present exist; or it is possible by itself and independently of the other differences because of something which exists continuously in order to make it possible, that is, it is implied in its hypothetical agent. We do not need both of these forms for that in virtue of which a feature of the organism is possible, and they are in contradiction with one another. And our choice between them depends entirely on whether we think that a latent quality is entirely independent of all other differences, and is a thing absolutely for itself, or whether we regard it as being conditioned by the other differences within the unity of the organism. And the biological treatment of latent qualities shows that, even in latency, the differences of the organism are looked on as *mere* differences, each of which is independent of the rest. Each belongs to its "vehicle," and is possible because of it. But it must be remembered that this solution of the matter is not so perfectly straightforward as it seems, for behind it there lies the question as to what are the grounds for the possibility of each " vehicle." This question is never answered, because the " vehicle," being unrelated, does not aspire to possibility; it is satisfied with abstract existence.

Now, this independent existence of latent differences meets with some difficulties. For there are very many of them. Transplant this primrose to Brazil, and it will probably change considerably; but it is difficult to think of those new qualities as existing independently in a latent condition in my garden. Am I covered with latent bruises because if you strike me I turn blue? If death, as they tell us, is a feature of organisms acquired by natural selection, then we may surely regard it as a latent feature existing in the individual independently of the rest by virtue of a special "vehicle" in the nucleus. How many latent tails has a lizard, and how many are my latent finger nails? For number plainly comes in when the possible objects are separate from one another. If a bone is broken and frequently disturbed, it will form a false joint at the breach, articular cartilage, synovial membrane and all; but we can hardly think of such a latent joint existing in a special vehicle at every part of a bone. And *all* the qualities of the *whole* organism are latent in a somatic cell which is capable of reproducing the whole. In fact there is much more *latent* in an organism than is ever actual at any one time, and if all these possibilities are separate things, we must invent a form for them in which they can be present in infinite numbers within a microscopic cell.

Again, certain latent differences do in fact accompany one another into existence. The manifold characteristics of an alternative form do not appear in independence of one another but together. There is every appearance of intimate relations between the various qualities of a sex; and if one of them exists the others also exist. In such a set of qualities, which may be proved by experiment to be one with one another, each one appears to imply all the others. And in not a few cases the nature of the *nexus* between them, that which is the identity for the whole set, may be more or less clearly defined. The "Evolution of Sex"[1] has nothing at all to do with evolution; it is a very brilliant essay on the physiological relatedness of morphological and other characteristics in the matter of sex. Latent differences again can hardly seem to be independent things, when one considers that the various parts in the successive stages of ontogeny are developed in intelligible and purposeful relations and in an intelligible order. In a word, the existence of latent qualities as abstract separate things in the form of vehicles of qualities is contrary both to probability and to experience. And it is worthy of re-

[1] Geddes and Thompson, Contemporary Science Series.

THE FIRST POSTULATE OF BIOLOGY. 81

mark that if they have existence as *Anlagen* at all, it must be as separate and discrete *Anlagen*. For if the differences have an immediate existence underived from the existence of one another but only derived from that of their respective anthropomorphic agents, there must be a vehicle for *every* difference; otherwise there would be latent qualities which existed only in possibility, because of one another, and not in abstract actuality in the vehicle; and if you deny self-sufficient existence to some differences, then why not regard them all as derivative and having their ground in the others? Seeing, then, that the theory gives us a form for the organism which is at once improbable and contrary to observation, what shall we regard as the impulse for this postulate in the matter of latent differences? Two points may be given in explanation of it—the definiteness of latent qualities on the one hand, and on the other that purely morphological consideration of them which we have already observed.

An individual becomes, let us say, male or female, or a shoot becomes a collection of green leaves or a flower, or a single element in the flower becomes a pistil or a petal, according to its nutritive conditions. Now, these were alternative latent characters in the various parts in question. And in such cases as these, if we were to substitute the definite conception of possibility for the indefinite conception of latency, we should immediately feel that the former was in some way inadequate to the case, and that it lacked something of the meaning of the latter. While admitting that the resulting image is in fact either male or female, either foliage leaves or flower, or either pistil or petal in every case, we should think that these alternatives can hardly exhaust what is possible to the stages before the various differentiations. Between these two alternatives how many forms might be intercalated, each of which is surely possible! How *possible* for the organism, if left to itself, and having within it no definite and separate representative for each quality that is to come, to err in almost

F

any direction! We do not talk of latent bruises, or of latent false joints, because these are not normal, and may occur anywhere; they, indeed, are possible and no more. But these definite purposive characters, which must be exactly so-and-so or be useless, are surely predetermined in some other and more distinct way than as the mere results of what we have given to us in the undifferentiated stage. Shall a difference in conditions which is simply quantitative, or at most, chemical, acting on an embryo which has not the two alternatives as two separate things now already within it, determine the creature to one of these two definite, complexly differing, yet perfect images? If the qualities are possible, they can only be so because the discrete exactness of each is laid down in discrete and special mechanisms. Otherwise the most various differences in conditions would have the most various effects on the image, and their results would not be restricted to the making of just this one or just that other of two, and no more than two, perfect forms. We have surely to do with a *mechanism* of such a kind that it transforms all effects coming upon it into the bare, punctual, momentary decision as to whether this pattern or that one shall be followed. The conception of possibility cannot be adequate to the expression of latency. This is, I suppose, the sort of movement which leads to the doctrine of the independence of latent qualities as separate things. And it is not easy to answer it.

This difficulty with regard to alternative qualities is little more than a special case of that problem of ontogeny which has usually been solved by the help of the same postulate. For the theories, the successive stages of the rise of the individual are separate things, which may or may not condition one another. They are usually unrelated, and if they are related then it is a "mystery." How, we are asked, shall this huge final definite system come out of that minute and comparatively homogeneous germ? The mass is vastly multiplied, and the process is not carried on under the ab-

THE FIRST POSTULATE OF BIOLOGY. 83

solute constancy of experimental conditions; it may even suffer considerable disturbances, yet the result is the definite adult. The germs of a thousand species are not distinguishable from one another; their conditions of nutrition and growth are so nearly the same that there is not room in these minute objects for much difference in chemical or structural respects, yet each one brings forth its kind. And this difficulty which we have in regard to the germ exists also, and in the same way, in regard to any succeeding stage. It amounts to this, that we cannot conceive of any kind of nature for the germ, or other stage, which should be such as to admit, with great exactitude, of the possibility of only one future definite image. We cannot conceive the germ as being thus the possibility of the image in virtue of those qualities in it which are known, or are such as might be known to cytological anatomy or physiology. To explain the future differences we assert, therefore, that they must be *there* already, in actuality, in the germ, each of them in virtue of its separate agent. I admit the difficulty, but deny the adequacy of the ordinary hypothetical solution of it. It does not seem possible that cytology should ever give us that difference between germs which corresponds to the difference between images; in other words, it does not seem possible that there should be an expression for the character of the whole in terms of an abstract qualitative, or quantitative, or morphological character of the germ. A mathematical, or chemical, or physiological formula for the character of a species is an attractive, but a vain dream. The one serious attempt that has been made to reduce the differences which exist primarily between sexes, and then between species, to terms of physiological tendencies—I mean that of Rolph, and later of Geddes and Thomson—was possible only because its main distinction was simple and unique. And a single distinction of anabolism and katabolism may give an idea of the whole difference of a gregarine from a flagellate, but it will not serve for the comparison of classes which are on different grades of organisation. Still less is it possible, as we shall see in the next chapter, to surmount the difficulty with the help of the anthropomorphic agent.

But if it is hard to believe that there is, in the mere natural qualities of the germ or of the later stage, and apart from the intervention of one-sided activity, the full determination of the possibility of development, the matter grows even more complex when it appears that the germ has alternative destinies, and that there are two possibilities. For these two, in the case of sex, seem to be without an inner relation to one another, just because they are externally so perfectly adapted. In proportion as we gain knowledge of the complex preparations of structure and function and feeling which are made in each for the other, so do these adaptations seem more intelligible as *for* one another within the species, and less intelligible and more arbitrary as *from* one another within the rise of the individual. Indeed, I hardly wonder that the latter aspect, as *ex hypothesi* arbitrary and unintelligible, has been given up to the "vehicle." For we appear to have two patterns which will not fit one another; and in such cases the tendency of men is to deny that one of them is an intelligible pattern, and to make it into miracle. In this case of alternative qualities the two patterns are, the alternative forms as purposefully complementary to one another on the one hand, and on the other the same forms as the two possibilities of development from an undifferentiated beginning. Now, if the germ, or the stage immediately antecedent to the critical differentiation stands (in its whole nature as for research and without agents) as the possibility of the one side, then the possibility of the other side must be wholly conditioned by that first possibility, and the two sides have an inner identity of origin, and are ontogenetically intelligible in relation to one another. But in that case the odds seem to be infinite that the outer identity of reciprocal adaptation would not be present in its large complexity and critical definiteness, and that the two sides would not be phylogenetically (as under the theories of organic evolution) intelligible in relation to one another. In a word, if you grant that the image arises by intelligible processes, and that it is not manufactured by the occasional intervention of the agent, it is difficult to find how the two sides should be *for* one another. If you grant that each sex cannot become so-and-so arbitrarily and without effect on the other in its rise, and that each stage of ontogeny cannot be so-and-so arbitrarily and apart from the rest, then *possibility* is your conception for the determination of the form of the individual, and the various differences of the organism are not without intelligible relation to one another. If, however, each alternative quality and each successive period of the organism exists in its own right, as a *latent thing*, then it is certainly easier to derive the form of the individual, under any formula of external usefulness which you may be in the habit of using, from the compounding together of such contrivances and parts and qualities as will make up a really good, working organism, adapted to its mate by special improvements, and fitted out with arrangements for almost all the circumstances of its life. It is the definiteness, complexity, and most of all the purposefulness of "latent" qualities, with the apparently inevitable consequent necessity of treating them as arbitrary with regard to their inner relations, which impel our theorists to bring them also under the first postulate of the hypothetical biology.

This postulate also rules the doctrines of organic evolution, in their one-sided method of finding the sole reason for every point of the organism in an often merely conjectural external use for it. It is easy to imagine the putting together of this and that adaptation in the course of a supposed phylogeny, because, in fancy, we may add anything to anything. Since each part exists in its own right, there is here no hindrance by considerations of possibility, and organic qualities which have admittedly no inner relations to one another may be combined or permuted, always in fancy, in almost any way you please. I say "almost," because it may be said that we have always this one guarantee of possibility, that the theories of organic evolution exclude forms which are unsuitably put together, with an external unsuitableness. We have, for instance, no animal with a skin notably too small for his body. And the conjecture as to why the parts harmonise externally is easily made. For the animal with a somewhat tight skin cannot, in times of plenty, eat so much as others do, or his peculiarity will repel rather than attract his possible mates, so that the ill-adapted form will not be multiplied. Or, by the continual pressure on his skin he will somewhat enlarge it, and the acquired quality of having a comfortable skin will be transmitted to his descendants; indeed, all three factors may cooperate. But such conjectures that organic parts will become, by imperceptible gradations, shifted, altered, eliminated and preserved by reason of their independent adaptation to one another, are very different from the knowledge that the possibility of the parts is in their inner identity alone. For the *former* restriction to theory merely demands, and not always successfully, that your suppositions should not be grotesque; but the *latter* demands that their form should not be such as to violate the possibility of morphology and of physiology. And the form of the doctrines of organic evolution implies that almost anything can exist with almost anything else in the organism. For it is not only based on the conception of independent and uncorrelated variation, but it works with the supposition that innumerable other forms, besides those of which we know, have existed and are possible. There would be little profit in debating that question, for it would all be conjecture from beginning to end; but I think that it might be possible to bring certain strong conjectures against

THE FIRST POSTULATE OF BIOLOGY. 87

the probability that many of the intermediate forms which our science postulates would be capable of life. In any case, however, this matter of unlimited conjecture from morphological homologies is unattractive, for who shall substantiate a word of it? And it is, more gravely, utterly untroubled by considerations of the rise and maintenance of the relations in the individual, looking on inheritance as the immediate reproduction of unrelated particulars.

Such, then, are the chief examples of the working of this first postulate, and it is hardly necessary further to point out its falsity. Organic differences of every kind are not separate elements, they are not numerable units, and the organism is not a mere sum of such units. To find that this is the case one has only to attempt to find one "character" in an organism which is not at once a part of a larger whole and itself capable of analysis into a hundred subordinate relations; one has only to consider that all differences of the organism arise together with one another in the development of the individual, and that the elements of morphology are *proportions* and that those of physiology are *functions* for the whole. The activity of every cell is not from itself alone, nor is the harmonious cooperation of all the cells an accidental or miraculous agreement in action on the part of discrete units. However much we may appear to gain for biology by separating the organism into things which play on one another externally, and whose particular activities can therefore not be intelligibly derived, we do no more, in reality, than to do away with the individuality of a natural system in order to invest its parts with the more unique character of moral agents. And the theory cannot even be called a working hypothesis, because it is impossible to define the limits of each unit, and because no one knows what is the nature of their agreement with one another. The manifoldness and intimacy of its relations with the environment and with other species and other individuals would make it impossible, except for purely formal hypotheses, to regard the organism as a total of elementary parts. For the relations of a living creature are infinite in extent; they go out into all the world. Yet there is no principle, under our postulate, by which we might deny to any part of this infinity the dignity of a separate adaptation separately represented in the germ. For everything that you can say about the body distinguishes one of its particulars, and it does not matter whether it is a distinction which is commonly made and considered characteristic or no. For if there are no inner relations for the particulars which are commonly noticed, it must also be allowed that the most searching or the most whimsical analysis can still find nothing but mere difference. In a word, the whole point of view is not one with which serious science can work at all. For science is just the continuous reduction of particulars, which at first appear unrelated and associated only by an invariable habit of occurrence, to terms of their inner identity; and if we are content to remain at the observation of invariable habits of concurrence, it is not easy to see how the mere recognition of the external usefulness of these invariable concomitances and sequences can carry our knowledge either far or surely. What we want is the middle term, as Dr. Stirling has it, which mediates between the extremes. For by it the special sciences of organisms do their work, and we may surely hope that it is not impossible that biology should recognise the rights of their method. This is to ask no more than that relations of reciprocity should be formally allowed for in our theories of the nature of the parts of organisms, and that we should not, in our science, utterly disregard that principle which rules in every other.

CHAPTER III.

THE SECOND POSTULATE OF BIOLOGY.

THE second postulate of biology is, that all the qualities of the organism and all its stages are the manifestation of, and are related to one another only through an agent or system of agents within the known body. The agent is purely self-determining, and is in a relation of pure activity to the body, which, in consequence, is in a relation of pure passivity to the agent. The latter may or may not be associated with particular microscopic appearances in the body, or with particular states of its feeling, but in all cases it is formally supposed to have such a phenomenal existence. It carries the qualities when they are latent, and therefore carries alternative qualities; and it manifests these when and where they ought to be manifested. If the agent is a "material vehicle of hereditary qualities" there is a distinct agent to each difference in the organism, but a system may be given to such agents by the hypothesis of a material "architecture" from which they are unfolded. If the agent is a quasi-psychical principle, there is only one, and its system is a system of ideas dimly apprehended.

The fallacy of the second postulate is, that the agent is merely the abstract principle of identity, to which a particular place has been given as one of and as among the differences for which it is identity; that as the product of a logical distinction hypostasized into a phenomenal difference, it is in contradiction with the possibility of research; and that it is essentially unknowable. It is found not to answer the problem, but to assume into its self-contradiction the distinction of categories which gives rise to the problem of the development of the individual. It is finally referable to the doctrine of abstract substance or of the thing in itself, and is found, like the first postulate, to be the mere application of the logic of abstract identity to the organism, in defiance of the methods of the observational sciences; so that we are again justified in regarding the systematic biology as an aberrant form of metaphysic.

The second postulate, or doctrine of agents, is the necessary consequence of the first postulate, or doctrine of the independence of qualities. For if any one quality has not a necessary and a possible origin in the others, and if the whole organism, as it may be known, is not the ground of explanation for each and all of its parts, there must plainly be a something of which all the appearances are the mere results. On this something they can have no effect in return, because to do so would be to have effects, though mediately, on one another. If one stage of development in the individual cannot find its explanation in what comes before and after it, but is cut off from the rest, as it were, with a knife, then it must plainly be the manifestation of something which, though it is not the organism, is yet in some way the unity of the organism. For there is in a living creature an overwhelming evidence of system; and as soon as we deny that the system is in the parts as well as made out of them, and make the system, so far as the parts are concerned, a mere external show, we must evidently bring forward some agent which itself shall carry the system in some other form within it, and reveal it in the otherwise unrelated parts. Borrowing a diagram which has been used for a different though a fundamentally identical argument,[1] we may illustrate the method as follows.

If $A\ B\ C\ D$ are successive stages of ontogeny, or contemporary qualities of the individual—it does not matter which—we may represent the ordinary method of research by the form

[1] Appearance and Reality.

THE SECOND POSTULATE OF BIOLOGY. 91

$$A - B - C - D$$

where the reason for each difference is found in all the others. In the various forms through which we have already traced it, the first postulate does away with this relation of inter-dependence, and finds that each stage or part has a separate, and so far absolute, existence. But the form

$$A\ B\ C\ D$$

cannot be regarded as anything but mere nonsense, though it has had its representatives. Our systems of biology therefore proceed to

$$\begin{array}{cccc} a & - b & - c & - d \\ \downarrow & \downarrow & \downarrow & \downarrow \\ A & B & C & D \end{array}$$

where $a\ b\ c\ d$ are the parts of a sytem of agents, and where $A\ B\ C\ D$ are merely derivative effects of these, as the one-sided direction of the arrows represents. For to give A any effect whatever on a, or to allow it in any way to condition the operations of a or b is manifestly to give up the whole point of view. For

$$\begin{array}{cccc} a & - b & - c & - d \\ | & | & | & | \\ A & B & C & D \end{array}$$

is manifestly no better than

$$aA - bB - cC - dD$$

which is our first form over again, making the agents quite superfluous and meaningless. And to invent, as some ingenious theorists of ontogeny do, the form

$$a - b - c - d$$
$$|\diagup|\diagup|\diagup|$$
$$A \quad B \quad C \quad D$$

is to have recourse to the very simple expedient of antedating one portion of any moment of the development, and of calling it the agent—a little juggle which I mention for the sake of completeness, but which need not detain us. Given this one-sided relation of the agent to the state or quality, which is determined by it, but which does not in turn condition it, I need hardly point out that a and b are unknowable. For if they could be known, they would be part of A and B. One could not describe a's action without implying in the same breath A's reaction, and so making a an element in A, which, for the purposes of the argument, it would be frivolous to distinguish from A. And further, if a and b are qualities or sets of qualities—that is, if they can be known—they are forever discrete and separate from one another, by the very argument which gave rise to them. Therefore any discovery about a and b would put them into the same category as A and B, and you would need to invent new determinant agents, and, for the purposes of the theory to point out that these—your new ones—were as yet beyond the power of the highest objectives. But I have said enough to show that the doctrine of agents, is, first, *made possible* only by the doctrine of the independence of parts, stages, qualities—in a word, of differences; and again, that the determinant agent is absolutely *demanded* by the independence and irrationality of the discrete particulars.

There are now, and always have been, two theories of the development of the individual. The first holds that for every difference in the adult image, there is a special difference in the germ, though the germ-difference need not be in the same kind as the image-difference to which it "corresponds." There have been two forms to this theory, one of them regarding the differences in the germ as exactly the same as the later manifested differences, so that the germ would contain the image in little; and the other regarding the germ-foundations or *Anlagen* for the qualities of the adult as differing in kind from the latter. Observation of the germ excluded the former alternative, but the latter is of such a kind that observation cannot touch it. For the *Anlagen* or germ-differences, each of which answers to a quality or part of the explicit organism, may be associated with any microscopical appearance you please, as a list of the theories would show. And as there is no reason for supposing that this granule corresponds with that quality, so there is no reason for supposing that this particular structure in the germ has *not* to do with that explicit particular of the organism, if you grant the method of the hypothesis to be valid. Further, since almost every organism may be determined by circumstances into one or two or more reciprocally exclusive and perhaps widely differing but perfectly definite images, as, to give the most general instance, in the case of sex, the germ-differences must be more numerous than the adult-differences, because the alternative qualities must exist in the germ together, as their corresponding germ-differences. And the number of these "vehicles of qualities" must be still more numerous than would yet appear. For we must remember that there are time-differences, successive stages, each with its host of qualities; and, indeed, whoever cares to go into the matter will find that if we give a "material vehicle" in the germ for every quality of the organism, the germ must be very large. But the differences in the germ need not be material particles, they might be the differences of an ideal system, quasi-psychically present to the germ; and the latter would determine the succeeding stage in conformity with ideas of the necessity of parts, and so on from stage to stage. These are the theories of the first kind, implying the existence in the germ of something which is not the qualities of the germ, but which *corresponds*, difference for difference, to the qualities of the germ and of every succeeding stage.

The second view is the "epigenetic" theory of the development of the individual. It is best defined negatively, with regard to the features of the first theory. It has no "vehicle of qualities" and knows nothing of a quasi-psychical principle. In a word, it is the critical point of view, and has no hypothesis. It may perhaps be distinguished from the other as the evolutionary theory, in so far as "evolution" implies the rise of what is complex and heterogeneous from what is comparatively homogeneous and simple, rather than the rise of a less complex system of differences from a more complex system of determinant "vehicles." By *possibility* the epigenetic theory does not signify, as the other does, *abstract actuality in another form*. The adult qualities are in the germ in possibility but not "in material representatives" or in a quasi-psychical knowledge, for everything that is in the germ is a quality of the germ itself. The epigenetic theory is thus the denial of the second postulate of the systematic biology.

Now, the biologists of hypothesis, as they subscribe to the first postulate, also with one accord subscribe, in one form or another, to the second. And some of them say, as they all imply, that the non-epigenetic view of the development of the individual is necessary to the doctrine of adaptation by the transformation of species. In this we shall find that they are right. The three postulates are closely interdependent. And because an hypothesis as to the nature of the determination of the form of the individual accompanies, as its complement, every hypothesis as to the manner of the transformation of species, it is necessary to pay that attention to these hypotheses of the germ-representation of qualities, which their frequency, elaboration and confessed importance demand.

Now, we have theories in this kind from many authors, but especially from Darwin and de Vries, Spencer, Haeckel and Wiesner, and from Naegeli and Weismann among the hypotheses of material vehicles; and from Stahl, Jaeger, and others among the hypotheses of quasi-psychical principles. The first class falls into three groups. The first of these consists of theories which give no hint as to how the "vehicles" come to do the right thing in the right time and place, and they differ further in including other qualities in the hypotheses than merely primarily morphological qualities. According to the third group, the "vehicle" is for a *part* of the organism, and carries all the qualities for that part; and its systems, especially those of Naegeli and Weismann, are very highly elaborated. Spencer and his legitimate followers find the secret of individuality in the external identity of minute constituent elements throughout the body, so that each of their units carries all the qualities of the body; but these theories, in their working out, forsake this form in all but appearance, and the unit becomes a determinant as contrasted with material which, to it, is purely passive. I shall shortly refer to five of these hypotheses, namely, those of de Vries, Spencer, Weismann, Naegeli, and Brooks, before proceeding to the general criticism.

DE VRIES' THEORY OF PANGENESIS.

This theory may be sufficiently indicated in the words of its author :—

" The hypothesis is one of intracellular pangenesis, and I shall give to the minute bodies, each of which represents one hereditary quality, a new name, and shall call them Pangenes.

" I consider the thesis, that all the hereditary foundations (*Anlagen*) of the organism must be represented in the cell nucleus, as the most important result of the last decade's research into the cell. I shall try to show that this thesis leads us to assume a transport of material particles which are the vehicles of the individual hereditary qualities. But this transport is not through the entire organism, nor even from one cell to another, but takes place within the limits of individual cells. The material vehicles of hereditary qualities are carried out of the nucleus to the organs of the cell. In the nucleus they are for the most part inactive, but in the rest of the organs of the cell they can become active. All the qualities are represented in the nucleus, but in the protoplasm of every cell only a limited number are represented" (P. 5.) And in another place: "This hypothesis leads to the following consequences. With the exception of those sorts of pangenes which are already active in the nuclei, as, for instance, those which rule nuclear division, all kinds must pass out from the nucleus, that they may become active. But most of the pangenes of every kind remain in the nuclei, where they multiply themselves, partly for the purpose of nuclear division, and partly for the sake of their distribution into the protoplasm. That distribution concerns in every case only those kinds of pangenes which must enter into function. In order to do so they can be transported by the currents of the protoplasm, and be carried into the proper organs of the cell. Here they unite with the pangenes which are already present, and multiply, and take on their activity. The whole protoplasm consists of such pangenes, which have at diverse times been emitted from the nucleus, and of their descendants. There is no other living foundation in the protoplasm." (P. 211.) And again, "The hereditary qualities must be grounded in living substance; every vegetative germ cell, and every fertilised ovum must contain in potential form within it all the factors which make up the character of the species in question. The visible phenomena of heredity are the manifestations of the qualities of very minute and invisible particles which are hidden in every living substance. And, indeed, one must assume *one special particle for each hereditary quality* in order to be able

THE SECOND POSTULATE OF BIOLOGY. 97

to give account of all the phenomena. I call these unities *pangenes*. Invisibly small, but still of quite another order than chemical molecules, each of them being made up of innumerable molecules, these pangenes must grow and multiply and distribute themselves in cell division into all, or at least nearly all, the cells of the organism. They are either inactive and latent, or active; but they can multiply when in either condition. They are specially inactive in the cells of the germ tracks, and they commonly reach their highest activity in the somatic cells. In the higher organisms it is probable that never all the pangenes in the same cell arrive at activity, but that in each cell one or a few groups of pangenes arrive at the mastery, and stamp their character upon the cell." (P. 188.)

I do not think it is necessary to quote any more of this theory of the qualities of the individual, or to describe its more highly developed parts, in order to make evident in it all the characteristics which I have enumerated as belonging to the systems of the biology of hypothesis. However full the theory may be of references to the facts of the organic world, it is yet not in the least supported by the latter, because the matter under discussion, the relation of qualities to individuality, involves only a feature of organisms which is universal and in respect of which organisms do not differ. It seems fairly evident that we have to do with a metaphysical question alone, in all these quotations, just the question of identity in difference, of substance and quality. The pangenes or anthropomorphic agents, each one of which is a material vehicle of a special quality, are unknowable, because if we knew them and found them in the body we should find the processes of each one affecting the qualities in charge of other pangenes just as much as its own quality, and we could hardly describe any one of these agents as anything but a set of qualities itself. They are anthropomorphic because they are purely self-determining and not passive, and because they know the right and do it. They

G

become functional *when it is time for them to do so*, they slip out of the nucleus *when they are needed* outside, they go through the cytoplasm to the part of the cell *which requires* their "quality." Their arrangements among themselves as to when each is to be active or latent, are carried on by conceptions of a normal order; for if it were only a blind struggle of each to express itself, carried on against the others and against the rest of the cell, then each would be conditioned by the others and by that which it is supposed to condition, and would simply be a part of the organism among other parts. And if that were the case it seems possible that even the quality which it carries would not be wholly independent of the other qualities.

MR. SPENCER'S THEORY OF PHYSIOLOGICAL UNITS.

This theory is in two respects the most serious and credible of all. The agents are at times represented by Mr. Spencer as being *similar to one another* throughout the individual; and as *constitutive* rather than *directing*, that is, as making up the bulk of the organism. I emphasise these points for obvious reasons. The first, if it were consistently held to, would secure a certain kind of identity present in and throughout all the bodily parts—an identity of material —in that a similar structure would be repeated in all the units; and the second point, if consistently held to, would do away with the uninspired remnant which is determined by the self-determining units, so that the whole body would be self-determined, and so that we should not be forced to look upon that impossible relation of the purely active to the purely passive of which I have spoken. The apparent value of the theory depends upon its apparent consistency on these two points. But each of them is not only vague, but is positively ambiguous for theoretic purposes. The agents are now similar to one another, and again dissimilar;

they are now merely constitutive, and again directing. The significance of this contradiction is plain. The question here, as in all such theories, is as to the relation of the universal and the particular within individuality, and Mr. Spencer's method is to make hypothetical units which are *either* identical with one another, thus supporting the universal, *or* different, thus answering to the particulars. Between two aspects of the individual he inserts his agents, which answer now to one and again to the other, but cannot answer to both; and the problem of individuality is answered by the self-contradiction of the agents. I quote some paragraphs which exhibit the strength and the fallacy of the theory.

"The tendency displayed by an animal organism as well as by each of its organs, to return to a state of integrity by the assimilation of new matter, when it has undergone the waste consequent on activity, is a tendency which is not manifestly deducible from first principles, though it appears to be in harmony with them. If in the blood there existed ready formed units exactly like in kind to those of which each organ consists, the sorting of these units, ending in the union of each kind with already existing groups of the same kind, would be merely a good example of Differentiation and Integration. It would be analogous to the process by which, from a mixed solution of salts, there are deposited segregated masses of these salts, in the shape of different crystals."

We have here a distinct statement of the difference of the units. They can be "sorted" and united with groups of the same kind.

"But as already said, though the selective assimilation by which the repair of organs is effected no doubt results in part from an action of this kind, which is consequent on the persistence of force, the facts cannot be thus wholly accounted for, since organs are in part made up of units that do not exist as such in the circulating fluids. The process becomes comprehensible, however, if it be shown that, as suggested in § 54, groups of compound units have a certain power of moulding adjacent fit materials into units of their own form."

Here we find that the agents are not merely constitutive, they are self-determining and other-determining; there is the agent and the material.

Mr. Spencer goes on to consider:—" If the compound molecules of the blood, or of an organism considered in the aggregate, have the power of moulding into their own type, the matters which they absorb as nutriment, and if, as Mr. Paget points out, they have the power, when their type has been changed by disease, of moulding all materials afterwards received into the modified type, may we not reasonably suspect that the more or less specialised molecules of each organ have, in like manner, the power of moulding the materials which the blood brings to them into similarly specialised molecules? The repair of a wasted tissue may therefore be considered as due to forces analogous to those by which a crystal reproduces its lost apex when placed in a solution like that from which it was formed. In either case, a mass of units of a given kind shows a power of integrating with itself diffused units of the same kind; the only difference being, that the organic mass of units arranges the diffused units into special compound forms before integrating them with itself. In the case of the crystal, this reintegration is ascribed to polarity —a power of whose nature we know nothing. Whatever be its nature, however, it appears probable that the power by which organs repair themselves from the nutritive matters circulating through them is of the same order."

And the hypothesis is supported by the regeneration of lost parts.

" If when the leg of a lizard has been amputated, there presently buds out the germ of a new one, which, passing through phases of development like those of the original leg, eventually assumes a like shape and structure, we assert nothing more than what we see when we assert that the organism as a whole exercises such power over the newly-forming limb as makes it a repetition of its predecessor. If a leg is produced where there was a leg, and a tail where there was a tail, we have no alternative but to conclude that the aggregate forces of the body control the formative processes going on in each part. And on contemplating these facts in connection with various kindred ones, there is suggested the hypothesis that the form of each species of organism is determined by a peculiarity in the constitution of its units—that these have a special structure in which they tend to arrange themselves, just as have the simpler units of inorganic matter.

The regeneration of the whole body from a detached somatic fragment, in the cases of *Begonia* and *Hydra*, is here described.

Mr. Spencer proceeds :—

"We have no alternative but to say that the living particles composing one of these fragments have an innate tendency to arrange themselves into the shape of the organism to which they belong. We must infer that a plant or animal of any species is made up of special units, in all of which there dwells the intrinsic aptitude to aggregate into the form of that species; just as in the atoms of a salt there dwells the intrinsic aptitude to crystallise in a particular way. It seems difficult to conceive that this can be so; but we see that it *is* so. For this property there is no fit term. If we accept the word polarity as a name for the force by which inorganic units are aggregated into a form peculiar to them, we may apply this word to the analogous force displayed by organic units. But, as above admitted, polarity, as ascribed to atoms, is but a name for something of which we are ignorant—a name for a hypothetical property which as much needs explanation as that which it is used to explain—nevertheless, in default of another word, we must employ this, taking care, however, to restrict its meaning." ("Principles of Biology—Waste and Repair.") And at a later place Mr. Spencer reviews the whole theory.

"Setting out with these physiological units, the existence of which various organic phenomena compel us to recognise, and the production of which the general law of evolution thus leads us to anticipate, we get an insight into the phenomena of Genesis, Heredity, and Variation. If each organism is built of certain of these highly plastic units peculiar to its species—units which slowly work towards an equilibrium of their complex polarities in producing an aggregate of the specific structure, and which are at the same time slowly modifiable by the reactions of this aggregate—we see why the multiplication of organisms proceeds in the several ways and with the various results which naturalists have observed. Heredity, as shown not only in the repetition of the specific structure, but in the repetition of ancestral deviations from it, becomes a matter of course; and it falls into unison with the fact that, in various simpler organisms, lost parts can be replaced, and that, in still simpler organisms, a fragment can develop into a whole." ("Principles of Biology," vol. i., p. 287.)

With these extracts before one, the self-contradiction on the point of identity is plain. The units throughout any one organism must be identical with one another (if the theory is to stand), with that abstract identity which absolutely excludes differences. For the same purpose they must be different from one another. It is hardly fair to lay any stress on the well-worn crystal metaphor, but it is plain that it requires identity throughout. But this is more evident when the theory is supported by the regeneration of lost parts. In that process we have the production of what is externally dissimilar to the rest. But the new part, because it completes the definite specific image, is produced by means of an inner identity with the rest. When a part has been removed, it grows again into the normal form, only because the units distinctive *of the individual* and not *of the part concerned* are the stuff for its repair. The lizard's tail is regenerated by *lizard units*, not by *tail units*, nor even, as I understand, by *lizard tail units*. Much more is the abstract identity of the units of an individual necessary to Mr. Spencer's exposition of reproduction from a somatic fragment. The bit of Begonia leaf is made up of, or contains, not particularised leaf units, but universal Begonia stuff; it seems, indeed, to be the identical units of the whole Begonia, merely appearing as leaf because of their circumstances, and ready, when these change, to become a whole plant of many differences. The substance for this theory must preserve its universality throughout intact. Otherwise, why base the theory on the reproduction of differences from an inner identity? If we have not identity in these agents, what have we?

But there is also an individuality of special organs, and the system is driven to give us organ units also. We have read of "units exactly like in kind to those of which each organ consists," and of a well-marked difference between organs as regards the structure or "polarity" of these elements. For organs repair themselves, and may reproduce themselves wholly by division. There are therefore leaf units, so distinguished not merely in virtue of their position and relations, but in virtue of their "polarity" or inner nature. These have therefore ceased to be Begonia universal units. But every organ is made up of a variety of tissues, which are themselves organs in as true a sense; these tissues are cellular; and the cells contain organs, nucleus, centrosome, envelope; nor are even these simple, but contain ordered parts which change and yet retain their special image. Has every system then, into which you may sub-divide and cross-divide the body, its own units of special "polarity," as well as the units of every system which includes it as a part, and finally the omnipresent units of the whole body? I think that the theory would not go so far as to say that. Therefore, if the physiological units are not identical throughout, answering to the universality of the whole body, and if they are not present in various grades, answering to every concrete universal, we will suppose them to be different in every part, answering to the differences of the body. In that case their differences are indeed great. My cartilage is more like the cartilage of the skate than it is like my epithelium. And I suppose that the same would be true of the respective units of these organs, and of their "polarities." And if that is the case, then what is the significance of the units at all?

Therefore, where the theory is merely vague, we must introduce a distinction. The author now treats the units as identical throughout the individual, and he necessarily does so. Again, and with the same necessity, he treats them as different. And we find that the body is not large enough to contain units answering to every difference and to every identity, so that our distinction cannot be a numerical one. The only possible form which remains to it is the logical form of universal and particular, the form of an ideal distinction. The units are different when considered in relation to the differences of the body, but they are identical when considered in relation to the ideal identity of these differences. When a distinction is thus substituted for the vague self-contradiction, the units themselves present that problem of the organism for the satisfaction of which they were invented. They have the two aspects of identity and difference, and can no longer be the identity for the given differences of the body, so that they become useless.

But, because the logical distinction is clear in the given body, these elements are introduced to hide it. They cover a distinction which is the subject of the problem by taking it into their double form of contradiction. We have no longer to ask what is the nature of this identity, which is given in the body, or what is its relation to the parts and qualities. We are told that those particulars are the manifestation of *(particular and different)* units, and that these same units *(being not different but identical with one another, and in virtue of that identity alone)* are the inner identity throughout the body which is necessary to the Begonia and lizard in question.

This is a good example of the method of explanation by hypothetical agents, showing how the latter, when analysed, imply a self-contradiction, and showing also that it is only in virtue of that self-contradiction that the hypothetical theory can even appear to answer the problem. Insist on either side of the contradiction alone, and the theory falls to the ground. Regard the universality and the particularity of the units as mere logical aspects of them, and it merely becomes superfluous, because it unnecessarily repeats the problems of the body which we know, in elements which we cannot know and of which there is no reason to suppose the existence. Use the units in one sentence as barely identical with one another and in the next as barely different, without turning the vague inconsistency into a distinction, and you have the most perfect of all modern theories as to the determination of the form of the individual.

But we have another difficulty in accepting Mr. Spencer's theory. Even if the units were identical throughout they would still fail to give us individuality. For they would still be different from one another with a numerical difference. And being thus parts of a whole, they must, in proportion as the whole is more truly individual, be more completely specialised. An endless repetition of similars cannot give individuality to the whole. As Mr. Bosanquet says: "A structure however complex which repeats itself homogeneously throughout all atoms of a certain substance tends to confer individuality, if at all, on the minute units in which the complex structure exists, but neither on the substance as such nor on its larger fragments; the supposed minute structure is not the structure *of it* or *of them*, but only a structure repeated *within* it or them." ("Logic," i., 138.)

And it is in this point that the great weakness of the familiar crystal metaphor lies. It is true that a crystalline substance comes out, with more or less water, into a definite form or into one of two or more forms. A crystal within its proper solution will also regenerate lost parts. But its constituent elements are similar to one another throughout, and their difference is only numerical. It is to the end no more truly one crystal than it is two or many. In these important respects it differs from the organic form. For the parts of the organism are, as we saw, heterogeneous, and their unity is the unity, not of substance, but of end. The only unity which we can find for the organism, and the only identity given in its differences is an ideal unity and identity, a kind of unity that is simply impossible with homogeneity of parts. Organic individuality implies the exact contrary to the identity of the elements of the crystal, with which it is so often compared.

WEISMANN'S THEORY OF IDIOPLASM.

In this theory, which is the most elaborate of all its class, we have a hypothetical mechanism for the "transmission of qualities" in inheritance, and for their working out in the development of the individual. There is a different agent in the germ for every part of the body, and the agent "carries" the qualities of the part to which it goes; it "controls" the cell which is its final resting-place. Further, the agents can remain inactive, during all the time in which they are being distributed, by cell division, to their respective proper fields of action. Weismann insists, and probably with reason, on the necessity of such a theory to a theory of adaptation by transformation. He says:—

"The germ-plasm is an extremely delicately formed

organic structure—a microcosm in the true sense of the word—in which each independently variable part present throughout ontogeny is represented by a vital particle, each of which again has its definite inherited position, structure, and rate of increase—*a theory of evolution appears to me to be only possible in this sense.*" . . . " All the more essential differences in the structure of organisms depend on this fact. The determinants are particles on whose nature that of the corresponding parts in the fully formed body depends, whether the latter consists of a single cell or of several or many cells. *The assumption of such particles is inevitable in a theory of heredity.*" ("Germplasm," p. 91.)

A considerable difference between this scheme and that of de Vries is already noticeable. Whereas the latter takes account of all qualities, and assigns to each its vehicle, Weismann's theory has chiefly to do with morphological qualities. His determinants go to cells, and each determinant undertakes the manifestation of all the qualities of its cell. The scheme is purely morphological, and takes no account of function, but we shall see that this abstraction from function is necessary to the method, and, indeed, is the more necessary the more perfectly the method is applied.

This germ-plasm is the "hereditary substance," and it is situated in the nucleus. It answers to the principle of identity in this case as in others, and this character is its real impulse. But its professed and explicit derivation is not convincing, and depends merely on the complexity of the arrangements of the chromatin within the nucleus, and upon the fact that this substance is halved very accurately (as, however, are all other structures in most cases) in karyokinetic division. The derivation is as follows:—

"We know that the nucleus contains a substance which, even with the imperfect means of observation at our disposal, is seen to be extremely complex, and that it becomes modified in a very remarkable manner after every cell division, only to be again transformed at the approach of the following division. We can, moreover, observe that the cell is provided with a special apparatus which evidently enables it to halve this substance very accurately. The statement that *this substance is the hereditary substance* can therefore hardly be considered as an hypothesis any longer." (P. 29.) This substance is therefore a complex apparatus for the transmission of morphological qualities and for the working out of them in ontogeny.

"According to our view, the power of transmission—which is possessed by all organisms, and on which the development of the higher organic forms is based—therefore depends on simple growth merely in the case of the very lowest conceivable organisms with which we are not acquainted; while in all forms which have already undergone differentiation, it results from the possession of a *special apparatus for transmission.*" (P. 466.)

And the nature of this apparatus is such that "*the germ cell contains at least as many determinants as there are different cells or groups of cells in the fully formed organism which are capable of being individually determined from the germ onward.*" (P. 467.)

"The histological character of every cell in a multicellular organism, including its rate and mode of division, is controlled by such a determinant. The germ-plasm does not, however, contain a special determinant for every cell; but cells of a similar kind, when, like the blood cells, they are not localised, may be represented by a single determinant in the germ-plasm. On the other hand, every cell, or group of cells, which is to remain independently variable, must be represented in the germ-plasm by a special determinant. Were this not the case, the cell in question could only vary in common with other cells which are controlled by the same determinant.

"The germ cell of a species must contain as many determinants as the organism has cells or groups of cells which are independently variable from the germ onwards, and these determinants must have a definite mutual arrangement in the germ-plasm, and must therefore constitute a definitely limited aggregate, or higher vital unit, the 'id.' (P. 452.)

" These determinants control the cell by breaking up into biophors which migrate into the cell body through the pores of the nuclear membrane, multiply there, arrange themselves according to the forces within them, and determine the histological structure of the cell. But they only do so after a certain definitely prescribed period of development, during which they reach the cell which they have to control." (P. 76.)

The nature of this "control" is as follows :—

"We suppose that the process in the idioplasm which brings about the ontogeny of a multicellular organism is due to the thousands of determinants, which constitute the germ-plasm of the fertilised ovum, becoming systematically separated into groups, and distributed among the successors of the egg-cell. This separation into smaller and smaller groups of determinants continues to take place, until each cell contains determinants of one sort only, and these then either control a single cell, or in case the hereditary character ('determinate') is constituted by a group of cells with a common origin, the control is exerted over this whole group."

And the whole matter is summed up thus :—

" The theory of heredity which has now been formulated, and more especially that portion of it which concerns the composition of the germ-plasm out of determinants, and the gradual disintegration of the mass of determinants in the germ-plasm during the course of ontogeny—is based on the assumption that the cells control themselves : that is to say, the fate of the cells is determined by forces situated within them, and not by external influences. The primary cells of the ectoderm and of the endoderm arise by the division of the fertilised egg-cell and its contained germ-plasm, because the determinants of the ectoderm are passed into one cell and those of the endoderm into the other, and not because some external influence, such as the force of gravity, affects the cells in a different manner. Similarly a certain cell in a subsequent embryonic stage does not give rise to a nerve, a muscle, or an epithelial-cell because it happens to be so situated as to be influenced by certain other cells in one way or another, but because it contains special determinants for nerve, muscle, or epithelial-cells." (P. 134.)

It is not necessary to go into any of the details of the theory. The *independence of parts* is essential to it. Each cell is controlled by its determinant agents; it is controlled, or determined, by "forces within it." Therefore cells are not in any way determined by surrounding cells, nor are the present states of cells determined by their former states. There is a complete abstraction from functional activity of parts, and even from such mechanical relations as stress and strain, gravitation, etc. Each cell is the *mere* work of the agent which controls it, and the successive stages of the organism appear to fly off from the system of agents—as a mere organism of appearance from a real organism—having no intelligible relations among themselves.

Now this theory gives us, as is always the case, agents which are used alternately as pure identity and pure difference for the body. But in this case we have an attempt at a picture of identity *in* difference. This is given in the "architecture" of the germ-plasm. The determinants for every part are in such a way fixed up together that they come off, during cell divisions, in the right directions, in the right kinds, and in the right proportions. Now this architecture is a system of differences, every one of which "answers to" a known difference. And the question arises whether we may not have in this theory something like what may reasonably be supposed to occur in the body. If the body persistently has a certain unity manifest in a host of orderly particulars, may the whole matter not be due to some such system of microscopic police?

Such a question might might be answered from research. We might base our denial on the fact that the division of the early embryo into halves gives rise, not to two separate half-images, but to two complete images each of which is of half the normal mass, except in so far as development seems to be disturbed in some cases by the abnormal want of material. Or we might base our denial on the familiar Begonia leaf of the arguments. A small portion of the leaf, we should say, has been "determined" by its own determinants, and contains only its own determinants, so how can it give rise to the whole plant? Or we might quote such an expert in the nucleus as Verworn, who, very naturally, finds that the changes in the nucleus are just as much referable to the changes in the cytoplasm as the latter are conditioned by the former. But such answers are hopeless. To the two first Weismann is ready to object that within the broken piece of the organism there is the whole police force, only that all of its members which are not concerned with that particular beat have been asleep, until an abnormal disruption calls them out. And to our third objection the answer would be that even Verworn cannot seize the peculiar activities of the determinants, and that *they* are unchanged by cytoplasmic changes. I merely bring forward these three objections from research to show how completely they are not to the point when we are dealing with the biology of hypothesis.

Our true answer is, that the action of the determinants, being one-sided, is unknowable, for the reason that, if it were known, it would be of such a kind that the determinant would be reciprocally conditioned by the determinee. Further, the conception of "inactive" determinants is a wholly false one. If they are there at all, they are there, and their influence counts for what it is worth. And if they are without effect on what is there, then they are simply not there. Do they not, even according to the hypothesis, assimilate, excrete, and multiply? And is that not activity, with a specific effect, according to their natures, on what is round them, and specifically conditioned, according to their nature, by the rest of the organism? But let us compare the determinant to an organism. Like the organism, the determinant can retain its proper form and functions, and is the same determinant through all changes. It is fed; it reproduces itself. It is not homogeneous, but contains many ordered differences, and in virtue of its qualities it does its work. Now, all its qualities, you will say, are surely not the mere results of one another, for if they were, it would not retain its identity through all the differences of its life any more than the organism would do, if cells were conditioned by cells and stages by stages. You, therefore, need another system of determinants to control the determinants of Weismann as soon as anything is known about these, and to be the vehicles of their qualities; and you must then examine that new system in order to see whether or no you need yet another.

NAEGELI'S THEORY OF IDIOPLASM.

In his great work, "Mechanisch-physiologische Theorie der Abstammungslehre," Naegeli elaborates a most ingenious theory of individuality and of the representation of qualities by agents. In the last chapter we saw his dexterous manipulation of qualities. He found for them, in their supposed existence in the germ, a condition in which they should be qualitatively similar, and yet in some way separate, and therefore numerically distinct and quantitatively measurable. We have now to see what the measurable form of the qualities is. "I think," says Naegeli, "of the qualities, organs, arrangements, and functions, all of which can only be perceived by us in a very complex form, as separated in the idioplasm into their actual elements; and the idioplasm brings into being the specific appearance which is peculiar

THE SECOND POSTULATE OF BIOLOGY. 113

to every organism by the necessary putting together of those elements." (P. 45.) Or again: "We must conceive of the idioplasm as bringing the foundations (*Anlagen*) for different organs into development, in like manner as the pianist brings, on his instrument, the successive harmonies and discords of a score into expression. He always strikes the same key for every *A*, and in the same way a special key for every other note. So, in the idioplasm, the groups of micellar rows lying side by side are like keys of which each represents another *elementary appearance* (unit quality). If, during the development of the individual, chlorophyll, or rather the chromogen from which, by the action of light, chlorophyll arises, is formed in any cell, the idioplasm there is setting the *chlorophyll-key* into activity; and in the same manner, when there are formed in a cell either spiral or dotted thickenings of the envelope, then the *spiral-thread-key* or the *dot-key* is being struck." (P. 44.) I think we may reasonably find in these expressions one old conception. That is, an agent which on the one side is pure difference—each of its differences answering to one difference in the known organism—and on the other side is pure identity, answering to the unity of the organism. I also think that in this one point we may reasonably find the sole impulse for and meaning of the theory, and that it is only in virtue of this—its fundamental—method that it can explain anything at all, and that it can explain everything which is organic. And thirdly, the undeveloped ambiguity of the agent in this, its only important respect, is necessary to the apparent adequacy of the method. But let us see how the theory works in particulars. The idioplasm is a highly complex structure which is distributed throughout the body in the form of a network. It forms part of the firmer portion of the protoplasm. It consists of fibres which in their turn are made up of subordinate units or micella. These ultimate units are microscopically invisible crystals, each of which consists of a greater or lesser number of molecules. "The

H

micella are most regularly arranged in the crystalloids of the albuminates, in some such way as the molecules are arranged in crystals; that is, in plane layers which cross each other in three or more directions; and also in straight rows of micella which are arranged in three or more planes crossing one another." The arrangement of layers and striation is recognised in starch granules, and in membranes of plant cells, where the visible layers split into two or more leaves. On this ground, Naegeli speaks of a *branching* of micellar layers or rows. Of such branching and growing fibres of micella is the idioplasm composed.

We know the structure of the idioplasm in one dimension only, that is, in the dimension answering to ontogeny. For the fibres grow to an almost unlimited extent in the development of the individual and in parthenogenetic reproduction. Wherever there is not a variation, there the cross-section of the idioplasmic fibre remains the same, and it only lengthens itself to the necessary extent for the individual and its descendants. Therefore, "the idioplasm is strictly arranged in parallel rows, which closely adhere to one another, but which grow by the interpolation of new micella, and thus continually maintain a constant arrangement." So long as this cross section of the fibre of parallel rows of micella remains the same, so long the qualities of the individual remain fundamentally, that is, as *Anlagen*, the same. But "when one or more longitudinal rows branch or unite because of a disturbance in their growth leading to a less firm connection between them, then the number of the longitudinal rows is increased or diminished, the configuration of the idioplasmic system is altered, and thereupon an alteration in qualities (*Merkmal*) arises. The constancy of the qualities during a series of generations is only possible when the micellar rows of the idioplasm maintain a strictly parallel relation throughout the process of ontogeny. And the alteration of qualities in the development of the race requires on the other hand an increase in the number or an alteration in the arrangement of the micellar rows, without which a new foundation cannot take its place in the idioplasmic system."

I shall not enter further into this quasi-histological hypothesis, which is fully elaborated with the aid of ingenious diagrams in the book. For all this part of the theory is merely there in order to tack on the agent to something which might be supposed to have, even though it has not, a phenomenal existence. It is more important to see what these *Anlagen* are in relation to qualities than to guess what they might be in relation to molecules. Now, the idioplasm is nothing but all the qualities reflected in another form; it is both appearances and possible appearances existing as actual things which are not appearances but are something else.

"Every perceptible quality is present in the idioplasm as a foundation (*Anlage*), and there are therefore as many different kinds of idioplasm as there are combinations of qualities. Every individual has arisen from an idioplasm which is somewhat differently fashioned, and in the same individual every organ and every portion of an organ owes its origin to a peculiar modification, or rather to a peculiar condition of the idioplasm. The idioplasm, which, at least during a certain stage of development, is distributed throughout the body, has therefore at every point a special peculiarity; in that it produces, for instance, here a branch, there a flower, a root, a foliage leaf, a petal, an anther, the foundation for a seed, a hair, a thorn." (P. 23.) The idioplasm, in fact, just consists of foundations, potentialities, *Anlagen*. And, where different appearances are, there are different "pecularities" of the idioplasm; difference for difference, it corresponds to the sum total of the known differences of the body, whether these are at the time in appearance or only in possibility. "The idioplasm of the germ is like the microcosmic copy of the macrocosmic or developed individual; for as the latter is made up of organs, systems of tissues, and cells, so is the former made up, in proportion, of hosts of micella which are bound together into higher units of different orders and represent the foundations for these cells, systems of tissues, and organs." (P. 26.) But the idioplasm is not merely in the germ; it is throughout the grown body. It is not merely in the germ that the qualities exist as separately represented and as numerable, although we saw in the last chapter how Naegeli goes to the germ to justify his conception of the independence of qualities. No difference in the body—but it has an idioplasmic difference to answer to it and to bring it, at the right time and place, into appearance. But this agent, like all the others, is not alone mere difference. Nor is it identity in difference, but is mere difference and mere identity confused. And we have to see on what principle its differences are arranged together within the unity of the organism. Naegeli is clear in his theory as to the conditions of the coming into appearance of the possible qualities as carried by the agent. The conditions stimulate the idioplasm, and it meets them with the *suitable* quality; and this conception leads to very remarkable qualities in the agent. For we find the idioplasm, during the development of the theory, driven at last to become a quasi-psychical principle.

With regard to the conditions of the development of this or that quality, Naegeli says: "In reproduction the organism inherits the totality of its qualities (*Eigenschaften*) as idioplasm. In the germ cell the qualities (*Merkmale*) of all its ancestors are enclosed as foundations (*Anlagen*). But the various foundations have a very disparate significance as regards their prospects of coming into development. For while some arrive at development always and without exception, others, under certain circumstances, remain undeveloped. In cases of the alternation of generations, for example, certain morphological and physiological qualities only appear in special generations, while they remain in the form of foundations (*Anlagezustand*) through a hundred successive generations. There are such qualities as only become actual under favourable outer conditions, remaining latent through whole æons (*Erdperioden*) because these influences are wanting. (Thus certain Alpine Hieraciæ show, when cultivated in gardens, a remarkable anomaly in respect of branching in the second or summer shoots of every year; in the Alps only spring shoots are formed, on account of the shortness of the time for vegetation, while the summer shoots never come into development.) Many foundations are reciprocally correlated or exclusive, so that the development of one of them either occasions or hinders the development of the other." (P. 24.)

Anything which is possible, although it may not appear for æons, is thus present as a separate foundation in the idioplasm. And it is worthy of notice that sometimes— even often—the actual presence of one quality conditions another quality, which is then in the relation of reciprocity to the first. In all cases where that is so, the "because" of one quality lies in the other, and there would be no need of the idioplasm, except that it is *through the idioplasm*, that is, through the unity of the organism, that the qualities in such cases affect one another; just as it is *through the idioplasm* alone that external conditions affect the coming into appearance of this or that quality. The agent, of its own movement, directs the apparent body, taking the grounds for its particular action from external conditions and from the qualities which it already has on show, but its movement is not explicable from those conditions and those qualities. It has left its quasi-histological derivation, and is no longer even *primum inter paria;* it is of a different order altogether from phenomenal qualities and changes. "The changes which the impulse to formation (*Bildungstrieb*) unfolds in the successive stages of the development of the individual and in the various parts of the individual, can be conditioned by nothing except by the successive modifications in the idioplasm, and by the likewise changing circumstances under which the idioplasm brings its foundations into development." (P. 32.) All that occurs in the individual is brought about by the idioplasm, which brings into light this or that quality out of its store of foundations. We are coming to see that the idioplasm *is* not that store of foundations, but that it *has* the store of possibilities. For if the possibilities, in this quasi-actual form of theirs, merely struggled for the right to appear, we should in the first place not have the order which Naegeli so truly makes *the* thing to be explained, and in the second place, we should not have that identity present in all the differences, which is necessary to the explanation of the organic form itself, and especially to the barest account of reproduction and of the regeneration of lost parts. The idioplasm has at first been described as a set of differences answering to the differences of the body. "It has at every point a special peculiarity; in that it produces, for instance, here a branch, there a flower, a root, a leaf, or a thorn." We are now to find it mere identity instead of mere difference. One author says:—"We cannot be surprised that the idioplasm is differently influenced in root, stem, and leaf; or that, notwithstanding its *perfect material identity*, the idioplasm brings to development, in these three different parts different foundations, but in their *lawful* order. We conceive that the nutritive conditions in general, although unable to make any qualitative alteration in the idioplasm, yet have an effect on the development of the foundations, inasmuch as, according to the quality and the quantity of the food, foundations may appear which would otherwise remain latent, and others which normally appear may be kept back." (P. 52.)

With this bare identity for the idioplasm, a somersault in theory which would be surprising if it had not been expected as necessary to the method, there comes in the possibility of an explanation of the orderliness and intelligibility of organic differences. The work of the idioplasm appears as "in lawful order," and the order is there because the idioplasm is one and identical. The agent is no longer a mere store of differences corresponding to phenomenal differences; it has become, by self-contradiction, materially the same under all the differences. It was a manifold, it is now a unity, in respect of quality. This is the essential self-contradiction which we remarked in Mr. Spencer's theory of physiological units. But now that the agent has taken on its character of identity, let us look at its functions. "The idioplasm," says Naegeli, "which is enclosed in the cells of given stages of ontogeny, directs the development of those foundations which *answer to the surrounding circumstances.*" All the foundations are there present, and the idioplasm takes account of the circumstances, and brings the *right Anlage* into appearance at the *right* time and in the *right* place; it undertakes the *lawful order* which we saw above. And this it can only do as one identity, if the thing is not done by the whole. It is not, after all, a group of foundations, but one foundation, and it was a mistake to suppose that it was different in different parts of the organism. And as a unity, the idioplasm is the origin of everything in the organism, yet not immediately the origin of everything after all. For "it is not necessary to assume that all molecular changes in the organism are separately impelled by the idioplasm; for in many cases that process would be limited to the setting of some of them in motion, whence a whole series of processes would follow in necessary consequences." Where "necessary consequence" can be found, plainly the agent is not needed. Its part is to do miracle, to interpose as *deus ex machina;* and from its actions whatever consequences are necessary will of course follow. Only, *its* action is the necessary consequence of nothing whatever. For the action of abstract identity upon the abstract differences is miracle, a miracle which has puzzled men in more sciences than one or two.

But the mere presence of a structural identity throughout the whole body, such as this network of micellar fibres bound together in parallel rows of which the cross section is always the same, is hardly adequate to the matter. For the one idioplasm rules the whole body, and a change *here* must be answered by another, and perhaps dissimilar, but always purposive, change *there*. Positively the network must be fitted with more than a mere histological identity through all its parts, in order to be a true principle of teleological activity. Every local change must be reflected, somehow, in the unity of the organism, and must be followed by activities which shall work the right and answering changes where they are needed. "Have we any guide," says the theorist, "and are there any instances in nature of a manifold which is ever reflected into unity?" The answer is close at hand. I quote Naegeli. "We may borrow an analogy from the movement which reproduces sense impressions and voluntary movements by the nerves. If the organised albuminates carry the most manifold perceptions of foreign objects in the finest degrees to the brain, and, producing there a clearly corresponding image, occasion in consequence answering movements, we may fairly assume that the organised albuminates of the idioplasm carry an image of their particular local changes to other parts of the organism, and there work out a change corresponding to the image. This theory of dynamic communication seems to me most simply to solve the questions which we have met. The idioplasm of all the cells of a plant is in immediate connection throughout. Every change which it suffers at any point will be perceived throughout and turned to account in a suitable manner. We must even assume that the stimulus which affects one point is immediately telegraphed over the whole body, and has at every point its just effect, for there is a continual balancing of the states of tension and of movement of the idioplasm. This continuous and universal feeling, which the idioplasm subserves, throws light on the otherwise astonishing fact that, in spite of the great dissimilarity of the conditions of nutrition and stimulation to which an organism is exposed in its various parts, the idioplasm yet retains, through all developments and changes, a perfectly identical form, as we see from the fact that the cells of the root, stem, and leaf, all produce the same individuals. The idioplasmic system of plants, which is common also to animals in the same manner, would thus be in many respects analogous to the nervous system. It would be, if I may so speak, a system of dynamic conductivity in a simpler and more material kind, while the nervous system gives us such a system of conductivity in a more complex and spiritual kind. Indeed, it is not unlikely that there is a phylogenetic connection between the two, in this respect, that in the animal kingdom one half of the idioplasmic system has by degrees become a nervous system." (P. 59.)

Thus does the idioplasm become a quasi-psychical principle; and the mere structural identity of the network throughout the body is found to be inadequate to the unity of the organism. And although the "dynamic conductivity" of the idioplasm is called "simple and material" in contrast to the more "spiritual" activity of nerves, I think it remains evident that the unity into which the manifold differences of the organism are reflected is, according to the paragraph, a unity of an essentially psychical kind. It is the *continuous and universal feeling* which is the secret of the *perfect identity* of the body; and by it "every change suffered at any point," that is, every difference, *is perceived throughout*. It is not necessary to study Naegeli's metaphor further than to make clear this point, that the idioplasm is like the nervous system, in respect that the latter is organic to the reflection of differences into the unity of conceptions, and that this point of resemblance is the only important and indeed essential one. It is in respect of this point of analogy that the idioplasm provides for the right and suitable reaction of this difference in answer to that difference. And because Naegeli's theory here passes essentially into the class of theories which deal with quasi-psychical agents, I shall leave the consideration of the unity of the organism as a unity of feeling to its proper place. But we may return for a moment to the analogy of the pianist, on page 113, and notice that it alone fully answers to this scheme. And it gives us, first, *the sounds*, answering to the manifold differences of appearance; then, *the keys*, answering to the idioplasm as mere differences; then, *the pianist*, answering to the idioplasm as abstract identity; and lastly, *the score*, answering to the ideal unity in multiplicity. Now, the analogy differs from the known body in one respect, that it inserts, between the phenomenal differences and the ideal unity, two steps; I mean, the abstract difference of the keys and the abstract identity of the pianist. And Naegeli's theory, like all other theories of ontogeny, exists only in order to insert those two steps.

I shall merely mention one more hypothesis in this kind, that of Brooks. This theory, which is a modification of the theory of pangenesis, begins by the clearest possible statement of our second postulate. Its author lays down the following foundation on its first page:—"The ovum is a cell which has gradually acquired a complicated organisation, and which contains material particles of some kind to correspond to each of the hereditary qualities of the organism." One could not better express the independence of organic qualities as numerable units; nor the carrying of each of them by an agent, which yet is not itself quality. The "*gradually acquired*"—very slowly, by insensible gradations, through long ages—is sufficient provision against any objections which might be brought from the rights of physiology, or indeed from any other point of view. Can this be included in the independent "rights of every science in its own sphere," that a theory should *begin* with a statement like the above postulate, as with a matter of common experience which admits of no questioning?

THEORIES OF QUASI-PSYCHICAL PRINCIPLES.

In his review of that metaphysical doctrine which we have found to be responsible for modern biology, Dr. Caird paraphrases one of its results as follows :—" We can conceive an external connection of things as acting upon each other; we can conceive a determination of that external connection by an intelligent being which uses it to realise some purpose or end ; but we cannot, according to Kant, form any definite conception of that, which yet seems to be set before us as a fact in organic beings, namely, of a unity which produces the differences of its parts and reveals itself in their determination by each other. In short, we cannot think of a unity that reveals itself in difference except as an intelligence ; and if we think of it as an intelligence, we cannot think of it as itself the source of the differences which it apprehends and on which it superinduces its unity, but only as an artist working with a given material." (Caird: "Critical Philosophy of Kant," vol. ii., p. 529.)

This way of thinking of the unity of a natural system is that which has given rise to the ever recurring and ever barren hypothesis of a quasi-psychical principle in the organism, as the unity for its manifold parts and periods. The multiplicity in question is an intelligible unity; it is therefore the work of an intelligent unit. The organic part subserves an end ; it is therefore fashioned, maintained, and prompted by one who is conscious of that end. Now, it would serve no purpose to enumerate the hypotheses in this kind from Bruno to Bunge, nor to point out their unessential differences, for the point of view is quite simple, and may be applied to anything organic in any way you like. But we must know why it is so fascinating to men and reappears so inevitably, and we must find out what it has to say for itself. Its most acute exponents are Bunge and Hartmann.

In his famous tract on Vitalism and Mechanism, the former lays down the familiar distinction between physical and chemical processes on the one hand, and vital processes on the other. Now, this antithesis is simply that between changes which are and those which are not referred to the individual or given in terms of the unity of the organism. Yet to our author these are not different aspects of one set of processes, but different processes which exist side by side in the creature, excluding and complementing one another. To him, therefore, there are changes and forces in the body which are not capable of the form of mechanical explanation, but which belong to a different sphere from that of dead causes and effects. And these other changes are quasi-psychical in origin.

Now, we may at once remark that we have here a certain confusion. The physical and chemical processes, as *mechanical*, are to be set over against psychical processes, *as in some way not mechanical*, but as free from reciprocity and as conditioned only by ends. But there is no reason at all for supposing that these two antitheses, between physical and psychical, and between what is mechanically and what is finally conditioned, do in any sense correspond to one another. We have no reason for excepting psychical *processes* from that form under which we include the rest of the organism. Thinking is not miracle, any more than "cerebration" is miracle, and, as a process, it is as much in bondage to necessity as anything else is. How then does Bunge expect that the *purposefulness* of the organic differences—a matter which *ex hypothesi* cannot be accounted for in terms of the physical processes in the parts—is to be accounted for in terms of the psychical processes in those same parts? Certainly the purposefulness is that which has to be explained, but the two kinds of processes which are here distinguished do not differ in respect of that matter, however much they may differ in respect of others. Both are, if both exist, equally purposive in fact and equally mechanical in derivation. And all that the theory seems to do is to add to one set of processes another set which does not at all help us in the explanation of the former.

Such an attempt to find the secret of system in the abstract activity of understanding agents is not uncommon in other spheres than that of biology. It is often assumed that the orderly articulation of society depends immediately on the well-informed activity of its government. Or its order is understood to exist because of the knowledge of that order in the minds of its citizens. Extreme cases of this fallacy are seen in the propagation of abstract systems for the form of society. But nothing could be further from the mark than all this. Social system is a matter for research; its intricate coordinations cannot be exhausted; we do not know it all, and *therefore* work it out. It is an intelligible system, but answers to no explicit system in an intelligence. So far is abstract understanding from regulating society, that theories of the state are wholly in bondage to the form of the state existing. Our doctrine of government is not the universal for all those particulars, it is not the identity revealed in them; it is itself only one particular among the others, and is not necessarily even an important one; our political ideas are among the stones, but they are not the architect. If we spoke of a mystery of society as we commonly do of a mystery of life, we should not succeed in tracing it to the functions of politicians who should entirely understand all the differences, and so determine them into an intelligible unity. We should not compare those functions with vital processes, as existing among and as the unity for the mechanical relations of those who are not politicians. Yet this is the point of view of the theorists of quasi-psychical principles.

"The mystery of life," says Bunge, "lies hidden in activity. But the idea of action has come to us, not as the result of sensory perceptions, but from self-observation, from the observation of the will, as it occurs in our consciousness and as it manifests itself to our internal sense. When our external senses meet with this same activity, we cannot recognise it. We see its results and accompaniments, the various forms of motion, but the thing itself we do not see. We have no organ for its perception, and we can only hypothetically accept its existence, which we do when we speak of 'active movement.' Every physiologist does so, and he cannot work without such a conception. This is the first attempt towards a *psychological explanation of all vital phenomena*. We transfer to the objects of our sensory perception, to the organs, tissue elements, and to every minute cell, something which we have acquired from our own consciousness." Certainly, "something from our own consciousness" is transferred, but that something is not psychical processes hypothetically inserted among the other processes in organic parts.

Conceive of an organism which has two parts, A and B. Now, A and B can each be derived from physical processes. We should be content with such a derivation if it were not that A needs and is needed by B, and that they are adapted to one another; and that the processes in the rise of each of these elements are so complex and so various that we cannot believe that the necessary end should have been so accurately reached through a path beset by so many chances, unless we suppose that there is in each of these elements another set of processes of a different kind. These hypothetical processes, then, in A, must be such as to affect and to guide the physical processes. They must also be such as to have a certain peculiar unity. For they are specially concerned with leading up A to its final form, so that they must include the conception of the later stages of A as end. And they are concerned with making A to fit B so that they must include the conception of the whole organism as end. Now whatever conceptions may be included in the psychical states of *A*, those conceptions are there because of its physical and other states; or, in other words, they are simply a part or a manifestation of its character, but are in no sense its character to the exclusion of the other states of *A*. There is no reason why *A*'s psychical processes should not affect its physical processes, but there is equally no reason why the latter should not as much and in the same way condition the former. In short, in so far as they have actions on one another—and no one-sided activity is conceivable—both sets of processes are together mechanical. Therefore the physiology of the hypothetical play of physical and psychical processes in *A*, if an account could be given of it, would destroy the value of the latter processes as principle of the teleological direction of the part. Let *A* think of *B*, and of its fitness for *B*, all the time; yet, *that* it thinks of these things is due to all its other parts and processes and functions, and not least to this, that it is already one with *B*. The psychical states of a part are, for the purpose in question, simply differences among other differences, and cannot abstractly furnish us with its identity.

It is evident that in thus attributing a quasi-psychical principle to each system, or to each part of a system, we fail of our object in two respects. We do not thereby gain an expression for the unity of the organism, and we increase its manifoldness. For the principle in question is nothing more than one part, the most important part, if you like, among others. There is nothing in the nature of psychical changes which should fit them to bring the multiplicity within which they exist into one. An intelligence is indeed an identity in difference, and it is perhaps natural that we should seek to insert such an intelligence into the organism, as the agent of its identity. But an intelligence is the unity of its own differences—its own states; there is no conceiv-

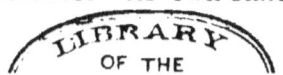

able sense in which it should be the unity for the parts of the body.

The whole method seems to depend on a confusion between *reason* of system and *understanding* of it, between the unity of idea and the unity of a thinking agent. There is, as Bunge says, something transferred from self-consciousness to the system without, and that something is also, as he says, certainly the origin of the conception of activity. But what is thus imposed on the outer world is nothing but the form of unity in multiplicity, it is no particular kind of multiplicity. The true anthropomorphism is the anthropomorphism of category and not of hypothesis. It is one thing to say that an organism is the product of reason, and another to say that it is the work of a thinking agent within it; the former is a metaphysical, and the latter a psychological statement. And since it is the mere form of organic life which is the ground for either, we can hardly avoid classing the psychological hypothesis with the doctrines of material vehicles, as a doctrine of anthropomorphic agents, and as infected with all the fallacies of that form of hypothesis. For there are no grounds for believing in the existence of the quasi-psychical principle, except such as are to be found in the intelligible form of the creature. It does not phenomenally distinguish itself from the rest of the body. It is, as Lotze has it, a homogeneous principle from which nothing can be derived, since we have no minor premisses to show why it should do this here, that there, and that other in yet another place. And it is hardly necessary to say that the attempts which are made to bring this quasi-psychical functioning into connection with the nervous system, or with the idioplasm, belong merely to the strange shifts to which theory is put in order to bring the quasi-phenomenal form of its agent into a pretence of probable relation with phenomena.

Such, then, are the theories of the determination of individual form which are given to us by the biology of hypothesis. Let us examine first their relation to fact, and then their general characteristics as a method. For the former purpose we cannot bring up against them particular facts which should disprove them, for their form is such that they will admit any fact. But they treat fact always in a certain manner, and we can observe whether their manner of explanation is at all satisfactory in regard to the wealth and variety of organic forms. We have already seen that, to them, all particulars are the same, and that the principle of identity is fundamentally the same; but it is conceivable that, in practice, this very formal treatment of fact might serve, if only provisionally, to unite different phenomena into intelligible schemes. Therefore, we try the theories in relation to the simplest animals.

There are, let us say, seven thousand species of Protozoa. The variety of their forms is very great. There are naked amœboid forms, and others with variously constructed simple or many-chambered shells of a siliceous, chitinous, or calcareous nature, or built of foreign objects which the animal has picked up. There is a vast number of spherical forms with exquisitely fine straight processes and siliceous skeletons of the most bewildering variety of plan. There are forms which swim freely by means of one or two flagella, or highly motile lashes, and there are others in which the body is variously supplied with cilia for purposes of locomotion, and of sweeping food in currents to the mouth. Altogether, there are not only very numerous species, but the differences between them are profound. Nothing can give the student any adequate conception of the variety and complexity of the Protozoa except a direct acquaintance with some of them. If he knows anything about their multitude, he will find that they have a peculiar significance in relation to the theories of biology. For he will ask what is the secret of their form, what is the inner difference between species answering to that outer difference of marks, and he will find that the customary "agent" deserts him where it seems to be most needed. And the reason is not difficult to find. It is, simply, that you cannot divide up these creatures, for theoretic purposes, into separate co-operating cells, nor regard their form as being due to a certain manner of cell divisions in ontogeny, nor regard their qualities as carried by "vehicles of qualities" in inheritance. You cannot, in short, in their case, delude yourself with the belief that individuality in organisms is a vain show due to the external action of an agent or system of agents upon the passive material which is known to us in research. Nor can you look at the changes of, say, a ciliate, and regard the morphological category as adequate to the expression of the differences between organisms. For structure, in such cases as these, is not there in its own right.

In the common process which is known as encystment, the individual loses its form, and becomes a spherical motionless mass, around which an envelope is excreted. Within this capsule it may divide into two or more masses, or into a great number of spores, or it may not divide at all. On their release from the capsule the practically amorphous individuals take on the specific form, sometimes in a few moments. To what extent every detail of form is lost in the process of encystment may be judged from the fact that two or more individuals may coalesce with one another, and form one spherical mass within the cyst. In many species it is possible to severely mutilate the animal with a knife, and the form will come again. In the division of a ciliate you may find the new mouth and new cilia arising in the posterior part of the lengthened body; and as a constriction gradually divides the whole into two individuals, the several organs for each one are separated from one another and perfected. In such cases as those which I have roughly described, the rise of form in the individual takes place, out of formlessness, before your eyes. And because that is so, it would be absurd to talk of the separate parts as carried by a mechanism for inheritance. Even if it does no more, the examination of the processes of change in the Protozoa destroys some of the most fatal refuges of ontogenetic theory. There is no conceivable kind of modifications of protoplasm which could be the ground of the essential difference between these seven thousand forms. There can be no talk of the "gradual" rise of structure in the individual by means of the orderly coming into appearance of separate *Anlagen*. These forms, at least, are truly individuals, and every part is *seen* to depend on every other. There is but one *Anlage*, and that *Anlage* is the whole, in every respect, and is no abstract chemical or other aspect of it. In a word, structure in such forms cannot be looked on as static, as we are accustomed to regard the structure of Metazoa. It is continually "becoming." And the very fact that in this group it is difficult, or impossible, to assign an external use for every part, brings into prominence that other fact, that there is an internal necessity for every part.

Consider the variety of the Foraminifera alone. I do not think any one would have the hardihood to say that every one of those geometrically regular forms has special advantages for the particular species, and *therefore*—by no other reason—exists. The very fact of their geometrical regularity points to necessities of structure rather than to arbitrarily combined advantages of structure; and we are not anxious to find several external places in nature for the species which range from *Uniloculina* to *Quinqueloculina*, just because there is *prima facie* evidence of the intelligibility of the actual and present formative processes which obtain in the individual. *Vorticellids* and similar forms may be exquisitely sculptured over their surfaces, and it is evident that this adornment at least has nothing to do with advantages to be gained by it, or with the attraction of mates, where there is hardly sex and no sight. The numerous species of the last named genus cannot be considered as having various respective advantages, in order to which each one exists. The exuberant variety of form in the Protozoa is quite inexplicable under the theories of organic evolution; the actual fashioning and maintenance of that form is untouched by the theories of ontogeny, and the theory of the independence of organic differences would require no further refutation than the mere mention of this group. I do not think that any biological hypothesis promises to throw any light on the determination of form in, say, the *Rhizopoda* or the *Radiolaria*. How is a body which is, for this purpose, structureless, and has nothing firm or rigid in it, whose substance is in a continually streaming movement, and which has, in short, no structure to begin with, to make this or that definite specific and complex skeleton? We do not know, and cannot guess at, the physiological processes of all this, any more than we can guess how a sponge makes its spicules or a holothurian its elaborately designed plates and hooks and anchors, or any more than we can give the most formal and bare account of the immediate determination of any organic form whatsoever. I believe that I am right in saying that no explanation of the immediate existence of any morphological element has ever been made. And this fact, veiled in the case of the Metazoa because in their case an external significance for the structure can so easily be found or feigned, lies open to us chiefly in the case of the unicellular animals, in which we are at once forced to see that form must have its *rationale* and to confess that this *rationale* is hidden from us.

Not only form, but also habit (of which form is but a function) lies before us in the Protozoa as a thing inexplicable by the hypotheses of biology, because we cannot in this group allow of the indirect action of advantage on structure or habit in phylogeny, as an adequate or sole reason for any character. There is too obviously an immediate reason, which, if it were known, would certainly profoundly affect and limit the explanation by external advantage; and the hypothetical intermediaries between the external advantage and the immediate presence of a character, which work out the character independently of the rest, are too obviously, in the case of these simple forms, not present. In a word, we cannot here, as we can in the multicellular animals, treat of a "character" as abstractly in the species, but we must treat it first as concretely in the individual. I do not deny that it is possible to insert such an intermediary even into the individual. The immanent soul does for theories of individual form and function what the factors of organic evolution do for theories of the species; they are both mechanisms by which the externally advantageous comes to be done and to be made and to be maintained, in complete independence of considerations of inner possibility and necessity, and thus in complete abstraction from real process. It is therefore barely possible to make external advantage the sole *rationale* of form and change, even in the Protozoa. I will give the stock example, from which it is commonly argued that the external advantage is secured by special processes existing *alongside* of the processes of mechanical necessity. Such cases as the following are of course not in any sense unique, but they serve their purpose particularly well because they are "marvellous." Bunge thus refers to Engelmann's researches into the movements of Arcella. "If a drop of water containing arcellæ be placed under the microscope, it often occurs that one of them falls on its back, as it were, *i.e.*, with the convex side downwards on the slide, so that the pseudopodia which appear at the edge of the shell cannot reach any support. It is then observed that, near the edge on one side, minute bubbles of gas make their appearance in the protoplasm; this side consequently becomes lighter, and floats up, so that the animal now rests upon the opposite sharp edge. It is now able, by means of its pseudopodia, to grasp the slide, and thus completely to turn over, so that all the pseudopodia are downwards. The gas bubbles now disappear, and the animal crawls away. If a little water containing arcellæ be dropped on the under side of a cover glass, and the latter is placed over a small air chamber, it is observed that the animalcules at first sink to the bottom of the drops. If they find nothing to lay hold of, large bubbles of gas are developed in the protoplasm, and, as they are thus rendered specifically lighter than the water, they rise in the drops. If they reach the surface of the glass in such a position that they cannot attach themselves to it by their pseudopodia, the gas bubbles are diminished on one side or increased on the other (sometimes simultaneously on both) until a tilting takes place, and the edge of the shell comes in contact with the glass, and they are thus enabled to turn over. When once this is accomplished, the bubbles again disappear, and the animal can now crawl freely about the glass. If the arcellæ are carefully detached by means of a needle, they at first fall to the bottom and then go through the same proceedings anew. Whatever attempt may be made to put them into an inconvenient position, they are always able, by the development of gas bubbles of appropriate size and at the proper spot, to right themselves, so that they acquire a position favourable to locomotion; and the attainment of this object is always followed by the disappearance of the bubbles. 'It cannot be denied,' says Engelmann, 'that these facts point to psychical processes in the protoplasm.' Whether this view of Engelmann's is justified or not, I do not venture to decide. I will even unreservedly admit that these remarkable phenomena may find a mechanical explanation. I have brought these facts to your notice merely in order to show you what complex manifestations of life we meet with, in cases where microscopical investigation has already reached its limit, and how little it has at present been possible to explain any single vital process on purely mechanical grounds." (Bunge: "Phys. Chem.," Engl. Trans., p. 9.)

A very great number of such examples are given by Hartmann, and are, for him, in every case the peculiar manifestations of the activity of the Unconscious. And, in the case of the Protozoa, changes which have manifestly an external advantage are frequently assigned to the special activity of a quasi-psychical principle which, *here and there*, interferes with an otherwise "mechanical" chain of processes. It is in this place especially significant, however, that there is no thought of regarding this habit of *arcella* as a separate "character," independently acquired, or of attributing its presence to the factors of organic evolution. The question has become, why *this* animal does it *to-day*? and the previous history of the species is evidently a matter impertinent to the question. Further, the supposed psychical changes are those of the whole individual, and we have here no really abstract agent which manages the various parts of the body independently of one another. And I believe that even de Vries would not be so consistent as to associate this habit of *arcella*, as one "character" among others, with a special material vehicle of its own. In a word, the theories of the hypothetical biology break down when one comes to apply them to the Protozoa, and the reason for this is sufficiently obvious. It has to do with the mere fact that the real changes which take place in them can be observed by the microscope, and that the building of their structure may be watched as it proceeds; everything is seen to be related within the individual, and to have its continual origin from the rest. That is merely to say, that we cannot, in their case, deny the fact of individuality, or resolve the animal into a multitude of independent cells, or into a combination of independent qualities; we do not therefore have recourse to the separate "vehicles of qualities," nor, in their turn, to the "factors of organic evolution." It is so evident, in the Protozoa, that everything which has an external significance has also, and first, an inner necessity and significance, that it is impossible to proceed with the explanation of any structure or change by means of its outer use alone, without making the attempt, at every step, to bring inner origin and outer use into one scheme.

Let us therefore review the chief characteristics of the hypothetical agents which are almost universal in biological theory. *They are not known and have not been observed.* This is evident from the extraordinary diversity and number of forms which they take, and from the fact that no one of the theories which deal with them can be either proved or refuted by means of an appeal to fact. *They are a scaffolding for the synthesis of abstract sciences.* Morphology, physiology, and embryology are studies which deal with aspects of the organism. If they are regarded as each containing truth which is complete in itself, and does not require to be corrected and fulfilled by the results of the other sciences, then we have facts of structure, facts of function, and facts of ontogeny, which are separate facts, existing side by side, and only connected with one another through the anthropomorphic agent. How is structure developed (since its functioning has nothing to do with its rise) but by the agent? How do the structures of the body complete one another and form an orderly whole but as the work of the agent? It is open to us either to criticise our abstractions, *or* to take them as answering to divisions in nature and to unite the separated aspects by means of the idioplasm or the immanent soul.

It is in virtue of that function that the agent is an *alogical principle*, the instrument for hiding and denying reason. There is no intelligible relation between those parts of the body, and between those stages and aspects of the body, and between those qualities which are externally held together by the agent. It is that which does things arbitrarily, so that one cannot know the process, and so that one cannot say what will be done next. It does—what it does; that is all. It is the bridge by which theory passes, in confusion, from the concrete to the abstract; it is the bridge from the individual to the species. Why has *this* body such and such parts? First, because the agent made them, and then because the agent has gradually acquired, in the course of many generations,—it is easy to imagine how—such and such habits. Structure is *for* its function; but its function cannot make structure, except through these hypothetical beings. In this way they are necessary to the theories of organic evolution. Why does this *arcella* blow its bubbles for hydrostatic purposes? It is a habit gained by natural selection? No; we do not care about the abstract habit; we can know nothing about that until we understand this case. Similarly, how has this dog those yellow spots above its eyes? It is a mark gained by natural selection, because it is an advantage to the dog to appear awake when it is asleep? No; we do not care about the abstract mark; we must first understand this concrete colouring. And for that you can offer us nothing but the agent.

The hypothetical agent was further found to be *unknowable*. It was unqualified and purely self-determined; there was nothing by which we should know it. Being separate from and among the parts over which it ruled, it was yet the identity which was present in all the differences of the body, and it was that which persisted unchanged through all the phenomenal changes. Thus it was both a mere particular among the other mere particulars, and yet was in itself the unity of the organism; and we found that it was these two things separately and for different purposes, but could not be both at once. It was the ground of the possibility of latent qualities as well as of those which are, at the time in question, in appearance; indeed, it was the actual existence of all possible things, as separate things.

In all these characteristics of the agent there is but one endeavour on the part of the theorists; it is, to find an expression for the unity of the organism. For the whole science is occupied simply with this matter of differences united in an identity, and with the nature of the bond of the members in a system. And the method which we have now observed is perhaps the simplest possible one, and it may be that some will consider that its two postulates are defensible as a working hypothesis. But it seems to me to be so riddled with contradictions as soon as it is taken seriously, and to be in any case so formal and inefficient, that we had better leave the whole problem alone than solve it by the empty doctrine of the independence of organic qualities and by the empty hypothesis of the anthropomorphic agent.

CHAPTER IV.

THE THIRD POSTULATE OF BIOLOGY.

"EVERYTHING organic exists only by reason of, and is to be explained only in relation to, some special external use which it now has, or which a similar structure has had in former times." This is the postulate for all the hypotheses as to the origin of adaptation, and as to the transformation of species. It not only thus underlies the greater part of the speculative biology, but it is also the foundation for the cruder forms of the doctrine of design.

Doubtless many consider this postulate to be a thing unquestionable, having a certainty of its own, and being necessarily the first principle for all biology. But its truth is incomplete and one-sided. When it is read as the biologists read it, it is as ridiculously inapt for science as are the statements that grass exists for cows to feed upon, or that the courses of the earth bring day and night in order that we may have light for work and darkness for sleep; in fact, it is just the same sort of proposition as these. Now, this third postulate of biology is the origin of those two which we have already considered, and of all the confusion which we have seen to arise from them.

Examples of its use are not necessary. We may characterise it as the very centre of such a theory as that of natural selection. For that theory, every part and every mark has its special use, which renders it necessary or at least advantageous, and thus, indirectly, favours its rise in the race. Under the form of sexual selection, again, every part or mark which can be brought under that form is found to have, and to exist only because of, its special and external value of beauty or attractiveness. And, in the scheme of Lamarck, we still find nothing else. He looks on every organic feature as being there because it was needed, and as indirectly taking its origin from that very need; and the particular feature is needed merely on account of its specific function or single use.

Our postulate thus finds, or conjectures with various degrees of probability, a "function" for every part and for every mark which can, on any principle, be distinguished from the rest of the creature; and this function, whether possible, supposed, or almost incredible, is held to be the sole *rationale* for the mark or part in question. Such features, on the other hand, as can be distinguished from the rest, and yet do not lend themselves to a special functional interpretation, are treated in one of three ways. Either their function is regarded as not yet discovered; or they are regarded as of necessity involved in the structure of other portions which do, in this respect, have functions; or, as we have already seen, they are interpreted as merely "vestigial" representatives of formerly functional homologues.

It may at once be noticed that, apart from other considerations, we are here led into a distinction between different kinds of organic features which is at once superficial and false. Some parts have special significances, others are necessary results of these, having no particular meanings of their own, and others yet, having neither special significances of their own, nor necessary interdependence with the rest, are merely inherited. We do ill thus to distinguish between three classes of parts or marks. Here are rather three aspects of all the parts of the individual, which have come to be regarded as the sole aspects, for science, of three kinds of parts in the individual. But we have reason for believing that everything in the individual may be interpreted in relation to a special function, that everything is also the necessary result and support of the rest, and that everything is also inherited. If then, we would do justice to the principle under consideration, we must regard everything organic as having its special use, and as existing only because of it.

The close interconnection between this postulate and the first and second hardly needs exhibiting. They all demand one another. If parts of a system are not intelligible in their inner relation to one another, then each must be maintained by an agent for a certain end. And if every quality has its special use, and originated from that use, then the chances are infinity to one that it did not also arise from and with the other qualities in reciprocal relation. We are near a very familiar question in this matter, but here is no place for its discussion. Only, the explanation by external use has often been held to be incompatible with that explanation which works with the universal and careless mechanism of nature. If that be the case, then our biology affords an apt example of it. For you must deny, by the first postulate, the validity and rights of the mechanical view of nature, before you can go on, in the third postulate, to affirm the unqualified pertinence of the teleological explanation. And conversely, after the dogmatic assertion of the independence of organic differences, one must go on to find a special and unique use for every one of them.

Now, this third postulate is by no means confined to biology. Its method and its peculiar defect may be observed in more than one other science. It has no small influence, for instance, in psychology, where the derivation of certain feelings or kinds of sensations is too often considered complete, when their utilities have been pointed out. And usually those utilities are conjectural in the highest degree. Fear, for instance, is there in order to carry the organism out of danger, or to paralyse him so that he may be taken for dead, and therefore uneatable, or that he may altogether elude observation. And when the argument has moved so far, it is easily switched on to the theory of organic evolution. Love is there, in order to add to our numbers, or, more subtly, to give rise to new variations, or, as others would have it, to eliminate and inhibit variation. The utility may be anything out of several contraries; we rarely know what it exactly is, but we know that there is a use, and are content to ascribe the origin and existence of the habit or feeling to that use unknown. And in so many cases is there certainly *no* biological utility in connection with a feature of our life, that recourse must usually be had to the "vestigial" form of explanation. Darwin's "Expression of the Emotions" has no other form. The scoffing smile uncovers the teeth for action, not because they are to be used immediately, but because it was formerly well to have them ready. And in the more acute modern science, we laugh in order to drive back blood to the head, which has been rendered temporarily anæmic by the sense of the ridiculous. Lunatics who haunt corners and avoid open spaces do so because their ancestors sneaked like cats from shelter to shelter in continual terror of death. Or they do so because those ancestors who did not hide were cut off before they had children. More likely still, there were positive advantages in the thickets, in that the berries and roots grew there, and there the females were to be found. Religion is unintelligible except in relation to its accidental uses, and, in despite of its essential unreasonableness, is there in order to the preservation of nations, which otherwise would fall into decay through too much intelligence. One begins to understand how it is that our modern quasi-biology can flourish through psychology and other sciences as soon as one grasps the simplicity and utter barrenness of its method. Every quality has its unknown use, and because of the advantage which that use confers upon the species, the quality has been established by the unknown processes of organic evolution. The method is inapt, and leads to nothing.

Proceeding upon this postulate, biology has to ask three questions: What is the special use of each thing? How do the parts exist by reason of their significances? And, How does the organism come to fit its environment? And the necessary limitations of the answers are as follows.

In regard to the first question, each part has an infinity of uses, all of which are, physiologically, as important to it as its special use, and very often much more so. In the consideration, therefore, of the origin of the part in the individual, and also of its origin in the race (in so far as the latter is considered to be the product of use-inheritance), the special use of the part may be of a minimal importance. Horns do not grow by butting, nor do anthers grow by fertilising, either in the individual or in the race. And innumerable other such instances might be adduced. The special use of a structure is thus defined by omitting the greater pàrt, and often the most important part, of its living relationships. I say most important, because we have to do with origins, and the special use of the part has not necessarily to do with origin.[1]

Now the special use is abstract, not merely because it is only one out of many, but also because it is often, from its external nature, more or less hopelessly conjectural. One cannot on that account say that there is not a use, for it must have an effect on the whole life, and that effect, whatever it is, can be looked upon as a use; but this peculiar nature of uses may render them inconvenient for science. What the effect is, is in many cases quite out of our knowledge. It might, often enough, be calculated if we knew a great deal about the manifold conditions of life. Even

---

[1] Still, even in the most unlikely cases, we may well be wrong on this point. Joints are for flexion, and false joints, in a fracture, are made by flexion. In view of that fact, can we go on to deny that normal joints are made by those appropriate bending movements for which they are made? Roux' work is full of interest in this direction.

then the research would remain a mere calculation, abstract and statistical. No more concrete approach could be made to it, so accidental is it. When we pass through such degrees of probability as the following, we see that any method of explanation which deals chiefly with the external uses of structures and marks, bases itself upon very uncertain data. The giraffe has a long neck to reach the leaves of trees, or, in drinking, to make up for the height of its shoulder. The electric organ of the skate is there in order to have some unknown effect upon its prey, which, however, it is not powerful enough to stun. Rabbits have white tails, that they may see one another in the twilight, or that foxes may better see them (one of these statements is just as sound as the other, and our decision between them could only be based upon some kind of statistical enquiry). Porcupines rattle their quills in order to warn their enemies as to what they are, and to make known that they are furnished with dangerous spines, so that they may not be attacked. In such cases—and you will find many even more doubtful in the works of the biologists—the use or external result for the life of the creature is practically outside of science, for this reason, if for no other, that chance enters so largely, and that our knowledge can enter so little, into its determination.

Some people are ever asking, 'Why do you do this and that?' They quite sincerely suppose that every action has its particular external use. Only in reference to that utility, and to the supposed isolated motive which springs from it, are they able to understand any action at all. To them your life must present the puzzling appearance of a set of disjointed actions, each of which only exists by reason of some end which is to be gained by it; and if the thing has no obvious motive at the beginning, or gains no distinct prize at the end, then it is a suspicious riddle. Now, these people are like the biologists. They deny the unity of character, and trace the teleological pattern of your life to an explicit conscious scheme for it, following the old path of the mere difference and the anthropomorphic agent, for man also can be studied, and wrongly studied, anthropomorphically. Giving a fictitious value to the results and to the uses of your action (which things are utterly accidental, and cannot be foreseen), they are ever labouring a hypothesis which should show some "reason," however remote and subtle, for what occurs. Your feeling, in giving reasons for your conduct, that you are deceiving, although unavoidably, is due to the abstractness of your answer, made necessary by the form of the question; you are giving a hypothesis, made at the moment, under special disadvantages, and though you have no means of knowing that there is any considerable correctness in your hypothesis, you know that it will be taken as having a value which it cannot have. The origins of the action are obscure and infinite, its results are infinite and incalculable, yet you are forced to patch up some kind of estimate of the latter, and to put it forward as the *rationale* of what has occurred. So impossible is the answer to the question, What is the special use of each thing? even in that sphere of human action of which we may be supposed to have more knowledge than we have of the animals and plants.

Who, for instance, would pretend to know the total results, however generally, of the special habit which Dr. Stirling thus discusses?

"Porcupines, we are told, rattle their quills and vibrate their tails when angered. These rattling quills, it appears, are only on the tail. Short, hollow, thin, open, supported on a slender footstalk each, they strike against each other and rattle when the tail is shaken. Mr. Darwin says further here:—'We can, I think, understand why porcupines have been provided with this special sound-producing instrument; they are nocturnal animals, and if they scented or heard a prowling beast of prey, it would be a great advantage to them in the dark to give warning to their enemy what they were, and that they were furnished with dangerous spines; they would thus escape being attacked.' It is curious how Mr. Darwin must always reason through the conjectural stories which he imaginatively gives himself to tell. But may we not also, equally imaginatively, conjecture some very different issue, or even a score of such? Even, as Mr. Darwin tells the story, it would be the 'enemy,' the 'beast of prey,' that would be advantaged—as warned not to make an attack where it would certainly only be injured. But to take it reverse-wise—it is, Mr Darwin tells us, an 'enemy' that is concerned. Well, what enemy that knew by its rattle where its prey was, and could come upon it by surprise and in the dark, would magnanimously consent to spare it till daylight, when it itself (the enemy) would necessarily have all against it which it had then and there for it? Really, when would Mr. Darwin wish us to suppose that this particular enemy seeks this particular prey? For, of course, the porcupine is like the rest, wholly in the drift of the struggle for life. We have to bear in mind, too, that, while the porcupine is in itself a very harmless, vegetable-feeding animal, it is only at night that its enemy is likely to fall in with it, for it is hidden asleep in its impregnably defended fortress during the day. The rattle 'a great advantage in the dark'! Why, but for the rattle, would it not be most likely to altogether escape its 'enemy' in the dark? And yet to Mr. Darwin it is precisely for this 'great advantage to them in the dark' that 'porcupines have been provided, through the modification of their protective spines, with this special sound-producing instrument,' an advantage which, as it turns out, can stead it in the second place only by steading its enemy infinitely more in the first place."[1]

All the results of anything organic are its uses. That we do not exhaustively know these uses does not invalidate them for the basis of our explanation of their origin or of their maintenance; but they do become useless for science

[1] Dr. Stirling, "Darwinism—Natural Selection."

by reason of the impossibility of knowing them at all in the greater number of cases, because of their accidental nature.

The second question—"How do parts exist by reason of their significance?"—is answered either by the theories of an immanent quasi-psychical principle in the individual, or by the theories of organic evolution. The objection to the first is that it is incapable of being filled by research, because it implies the hypothetical agent which we have already studied. And the theories of organic evolution have very much the same significance. They received their origin from a determined attempt to eliminate what is known as *design*, from nature, and their curious mechanism is to be referred entirely to that endeavour.

They consist of various schemes by which the purposeful event may appear to be the result, purposefulness and all, of a tumble of circumstances which are not concretely intelligible in relation to it. Or, as the theorists themselves would put it, they strive to show the rise of the adapted from " merely mechanical processes." " The old argument from design in nature fails, now that the law of natural selection has been discovered. There seems to be no more design in the variability of organic beings than in the course which the wind blows." These words of Darwin roughly indicate the impulse, and the secret of the attractiveness, of these theories. It must be remembered that the latter usually oppose themselves to an exceedingly crude and elementary conception of design, and rightly refuse to give much importance to a doctrine of the formation of all organic forms by a mere unknowable agent. For in such a doctrine we have no more than the old quasi-psychological form for an ideal unity.

Parts, therefore, though existing by reason of their significance, do not exist because of someone's perception of their significance. Now our theories, in their abstraction from the physiological reciprocity of parts both in the rise and maintenance of the individual, indicate that the alternative views of the parts, as existing because of their significance, and as existing because of their conditions, respectively exclude one another. They therefore attempt a scheme by which these two aspects of phenomena may be *historically* separated from one another. Conditions, we are told, do not directly affect parts, but the latter are suited to conditions by an indirect process of progressive adaptation in organic evolution. In such an argument we must remember that the concrete and particular parts and conditions which exist in *this* organic relation on which I place my finger are not introduced at all. "The heart and the pericardium" are by no means the same as this heart and this pericardium, and though you may invent what origin you like for the relation between the former abstractions, you are not at all touching the concrete problem. But parts are immediately freed from their conditions, in their rise and maintenance, by the insertion of the idioplasm or of some other such principle, or by the formal allowing for it. And the secular modification of parts by conditions is, on its side, only possible through such an agent.

Now, this logical feat on the part of the theories of adaptation by organic evolution is the beginning and end of their existence. If it were not for their metaphysical task and trick, they would not continue to exist for a day. For one may see how utterly without content they are by considering the ordinary biological controversies, in which the most opposite conclusions may be both impregnably and ineffectually held. And what, in this way, they can do for organisms, that they can also naturally do for not a few other systems in which the cause and the end seem to be strangers to one another.

It is the merest affectation to deny that their uses for the life of the individual must in some way enter into our explanations of the rise of parts both in the individual and in the race. Eyes are made for seeing, and are maintained by seeing, and joints are made for and are maintained by flexion. And in just this way everything organic appears to be adapted, as soon as you can find the point to which it should be referred. The problem of biology is to bring this *made for* and that *made by* into some intelligible union. But the Darwinian doctrines do not effect such an end. For their form, and that of all which have sprung from them, are as follows. In the individual, the part is intelligibly *made for* its function, but is unintelligibly *made by* the agent. In the race, on the other hand, we are to know that the part is selected or is used by its function, but is not *made for* anything at all (for here, at the great distance of "fortuitous" variation, mechanical process is allowed to come in). In other words, we have here an arrangement ingeniously planned, by which we may pass without a jolt from the category of ends to the category of cause—a mere mechanism for a formal translation. By it we are enabled to give a formal account of the origin of parts, under the conceptions and categories of physics, while we really and in practice treat of their origin as determined by the prescience of their ends. Ask our present question again, and you will see that it is the problem of the greater part of the hypothetical biology; it alone impelled Lamarck and Darwin, and impels Weismann and his critics. And you will see that there are two ways of bringing together the existence and the significance which form its terms; first, by introducing a whole world of hypothetical agents and processes between the terms; and second, by the study of the relation of these two aspects of everything organic, based on the recognition that the significance and the existence go ever hand in hand. In either case we have to do with a logic, and the only difference is that in the former case we have unverifiable hypotheses, but in the latter case we have a study of the methods of research. Everything organic certainly exists, as our postulate would have it, by reason of its special significance to the individual. But it does not, in fact do so through all the Darwinian or other apparatus, but in some way even here and now, because it is organic to an individual.

The third question is only a form of the second, but it is perhaps the most common form for biological speculation. In attempting an answer to it, we are to remember that organisms are, in a sense, not adapted to their environments. That is true in so far as the word "environment" denotes facts and conditions which are not *their* facts and conditions. It is true, inasmuch as we can easily think of a thousand improvements by which they would attain special advantages, but which they have not. But a creature's environment cannot be studied apart from the study of its structure and its habits. If we do so, we are describing something to which the organism is plainly not related, and to which it is consequently not adapted. But when we mean, by "environment," the known relations of the organism, then the latter is wholly adapted to the former. If a structure appears to fail of its end, we are to learn, not that adaptation is as yet incomplete, or that it has fallen off from perfection, but that we have not yet taken hold upon the true use of the structure. This statement is incomplete, but it is not incomplete as regards the doctrines of organic evolution. For the environment is just as much a part of the creature as are its morphological proportions, and in the study of adaptation we have no right method of distinguishing between the relations of organs within a body, and the relations of parts of the body to what is called "without," as between two different kinds of relations. Are we to speak of the weight of the organism, or of its environment of gravitation? Is an animal slow, or is its enemy swift? At what point, or on what principle, does the oxygen and the moisture of the air cease to be mine and become that of the environment? Or, in speaking of the skin, can we find any essential difference between the kind of its relations to that which is beneath it, and the kind of its relations to what is above? Is not the whole body environment, and the whole of its conditions body?

In this matter we come to the old difficulty, that biology is drawing hard divisions where research knows of none. For it is obvious that we cannot speak of a relation where we have not two sides which already act upon one another. Research knows nothing of one outer side, to which there might be an adaptation, if the inner side will only some day change. The relation, in short, is either there or it is not there, and if the latter, then it is nothing at all. But biology gives us an outer to which the inner is not yet, but will some day be, adapted. In so far as the outer acts upon the inner, it is one with it. And in so far as it does not act upon it, it is nothing relevant to the matter. A man, for instance, undertakes severe exertions, and his heart is hypertrophied. He is living, let us say, in a hilly country, or is occupied as a hunter. The environment, in this case, is the hills, and the adaptation is the strengthening of the heart. Now, in such a case we have an intelligible relation, which can be worked out in every step, between first and last. But let him live on the plains, and never go up the hills, and the latter, however high, will make no difference to his heart. The environment, in fact, is not his, except in virtue of its being physiologically a part of his life. The reaction, in this simple case, but also, we must believe, in all cases, is a reaction which, both in its manner and in its limits, can be followed in detail.

When, however, we come to organic evolution, we find that there is either no reaction, but a mere happening together, such as we have in the theory of the selection of fortuitous variations, or there is a reaction which is confessedly unintelligible, such as we have in the theory of use inheritance. And in both cases we, therefore, reach the possibility of explaining the becoming adapted of the organism to an environment which is not *its* environment. The species becomes fitted to conditions which do not immediately condition it, by steps which are undiscoverable, and which do not admit of being even guessed in their detail. It follows that adaptive changes become a set of changes which are of their own kind, and stand apart from the changes of intelligible reaction, the latter sometimes accidentally taking their place among the former, but never doing so otherwise than accidentally.

It is not necessary to examine the factors which have been brought forward to supplement the action of the two great principles which divide the theory of Lamarck from that of Darwin. Nor is it relevant to the question, to adduce observations from organic life. For all theories in this kind abstract from the processes of ontogeny, and this abstraction appears to me to be quite fatal to their claims. Throughout their whole extent, we are to remember, they are not dealing with anything which happens or which can be observed. It is for the reader to judge what disadvantages such a method confers upon theory.

LAMARCK.

There are reasons why Lamarck's system should be of special interest to the student of biology. For it contains the first theory of adaptation through the transformation of species, which is founded upon bare hypothesis, and it may in this respect be regarded as the pattern for all modern biology. It also gives its name to that school which somewhat vaguely opposes itself to the explanation of everything organic as the product of natural selection and fortuitous variation. And thirdly, the hypothetical processes and substances of this system have now become, in the ordinary course of research, obvious fictions; so that the doctrine reveals at once its true relation to observation, even without the demonstration of the impossibility and self-contradiction of its typical hypotheses. For Lamarck continually boasts of the sure method of observation, and fills his pages with innumerable references to organic phenomena; but the advance of research has shown that he, like the other biologists of hypotheses, was deceiving himself on this score. For, whatever had been the actual qualities of organisms, they would have supported his theory equally well, so long as they were qualities of organisms.

But even apart from the light which the study of Lamarck throws upon biological problems and methods in general, the system itself is of great interest. It is certain that this biologist has been, on the whole, misinterpreted, but, on the other hand, it is not easy to be sure that one understands him. And the confusion has taken place chiefly over that key-word of his system, *besoin*, or need This word has frequently been rendered "desire," and Lamarck has often more than a trace of this meaning in his use of it. But, on reading the whole development of his argument, one immediately finds that neither interpretation may be used exclusively, and that, in fact, *besoin* is a conception which, for the purposes of the theory, must remain undeveloped and undefined. Use the conception of need alone, and the theory is inadequate; restrict yourself to the conception of desire, and the whole theory becomes ridiculous, as flippant controversialists have not been slow to find. It is certainly more than need, and as certainly, it is less than desire. It probably appeared to Lamarck to have a quite definite meaning, requiring no further analysis. One can only suppose that he was unconscious of his easy bridge from the ideal relation of necessity to the psychological fact of desire—from the logical form to the phenomenal process.

In the need, or *besoin*, we have not, of course, to do with a mere negation. It is not a mere being without or not having. The organism has some kind of reference to the thing which is needed. And, since every part of an organism is needed by the rest and by circumstances, it is not difficult to find, in this *need*, a general principle for the origin of all organic characteristics. All characteristics are alike in this point, though in no other; for simply as parts of organisms, they have relations of necessity with one another. But a mechanism, by which this form of necessity may create the particulars, is wanting to observation. Yet such a mechanism becomes desirable, in order that the emptiness of the theory may be, if only apparently, filled up. The logical distinction must therefore, as we are well accustomed to find, become a quasi-phenomenal difference; and thus there arises the Lamarckian desire, which is more like Hartmann's Unconscious than it is like anything else, being quite as metaphysical in its origin as is the latter. I shall not attempt the difficult historical question as to what Lamarck supposed his conception of need to include. For our present purpose I merely take his work as it stands, and study it in relation to the problem of adaptation, and watch its development of the old quasi-psychical principle. And I here set down certain important passages from the " Philosophie Zoologique," chapter vii.

"It is evident that the observed form of animals is the product, on the one hand, of the ever-increasing complexity of organisation, which tends to form a regular gradation ; and, on the other hand, of the influences of a multitude of very various circumstances, which tend continually to destroy the regularity of that gradation of the increasing complexity of organisation. But I must explain my meaning in these expressions. 'Circumstances affect the form and the organisation of animals' means that the former, in becoming very different, change, in course of time, even the form and organisation by proportionate modifications. Certainly it would be a mistake to use these words literally, for, whatever be the circumstances, they have no directly modifying effect whatever on form and organisation. But great changes in circumstances give rise to great changes in the needs of animals, and such changes in their needs necessarily give rise to changes in their actions. And if the

THE THIRD POSTULATE OF BIOLOGY. 155

new needs become constant or very lasting, the animals take on new habits which are as lasting as the needs which gave rise to them. It is easy to demonstrate this; indeed, it is obvious without explanation. So that it is evident that a great change in circumstances, when it has become constant for a race of animals, leads those animals into new habits. And, if the new circumstances which have become permanent for a race of animals have given rise to new habits in them (that is, have impelled them to new actions which have become habitual), the result will be the use of such a part in preference to that of such another, and in certain cases the total disuse of such a part as has become useless. Now, none of this is hypothesis or my own opinion; it is, on the contrary, truth which only needs attention and observation of facts to become evident." (P. 222.)

Such, then, is the process of adaptation, to some extent in the individual, and altogether in the race. And the process is the same in each case. Moreover, the same process might be thought to underlie the adaptation of part to part within the development of the individual. A race of animals comes into certain circumstances, or is in them (for the change is not essential to the argument), and suffers no direct change from these circumstances. The latter, however, affect the needs, and the needs condition the actions, which, as habits, affect the form and organisation. Just so in the theory of natural selection, the circumstances do not directly affect the race, but all adaptation is referred to the indirect affection of form by the environment. This indirectness of the relation of the individual to circumstance is the form which is first attained by all theories of adaptation, and the special manner of it is a secondary point.

The meaning of this indirectness is, that no particular of the organism is changed, except through the unity of the organism, and that changes in organisms have the appearance of being purposeful responses to, rather than mere results of, the changes in the circumstances. And the theories which we are studying have no other object than to derive the present purposeful reaction from a series of direct causes which operated long ago. But it must be noticed that this element of theory, which is as strongly held by Weismann as by Lamarck, completely does away with all those analogies for organic adaptation which are drawn from inorganic things. You may see the coat of rust round a ball of iron, or the shore round a bay, given as parallels to the adaptation of one part of the organism to another, or of the whole to its environment. In such cases the relation is perfectly direct, and the change in one element involves a calculable complementary change in the other. But it would appear as though such a direct action of circumstances on organic form takes place, if at all, only within the narrowest limits, either in the individual or in the race. We have seen that the nutritive conditions of form are not answered in the individual by results such as one would expect from the conditions, or such as one could give in parallel degrees with the conditions. The determination of sex, for instance, is hardly the direct result of the food supply, in the same sense as the shape of the bay is the direct result of the shape of its shore, and the difference does not *seem* to be merely that of the extent of our knowledge. Let a race of birds take to the water, and, provided that such changes do, in truth, occur, its toes become webbed and it secretes oil for its feathers. But these changes are hardly the direct result of cold and wet. In the great majority of cases, one can see the advantage of a structure, and when its advantage is not evident, it is looked for as probably discoverable. But either one cannot speak of an immediate cause at all, or one cannot relate that cause to the advantage. This is notably the case in the study of ontogeny. The events which produce muscle have apparently nothing to do with contraction; bone is not, apparently, produced by stress, nor, so far as one can see, does nerve arise in the individual by feeling. Still, this may be only apparently so. Of course embryonic parts have functions, and it is probable that these are not incomparable to their adult functions. But, in fact, we are accustomed to consider rather the end than the cause of the organic part; and when we attempt to find the end as cause, it is plain that the end must operate indirectly, for there are other immediate causes which can be found by research. Hence there arises the necessity of making adaptation indirect, and Lamarck, as we have seen, is plain on this point. What qualities the organism has, it has because it needs them and because they are purposive; for the form is not immediately referable to the circumstances in which it finds itself. That is the beginning and end of the theory, and the rest is only the very vague and hesitating attempt to show how the need can bring about the new structure. That this is Lamarck's central thought appears from his summary of the theory, which includes his two well-known laws. He says, "In order to see the true order of things one must recognise :—

"(1) That every change which is at all considerable and continuously maintained in the circumstances of each race of animals, affects in it a real change in their needs.

"(2) That every change in the needs of animals necessitates other actions on their part for the satisfaction of the new needs, and, in consequence, other habits.

"(3) That since every new need requires new actions to satisfy it, it demands of the animal which experiences it either the more frequent use of such a part as was formerly less used, so that it becomes considerably developed and enlarged; or the use of new parts which insensibly arise in the organism from the needs, *by the efforts of its inner feeling*, as I shall presently show from known facts. And so, to arrive at the true cause of so many different forms and so many various habits as are given in the animal world, one must recognise that the infinitely diversified but slowly changing circumstances in which the animals of each race have successively been placed, have brought about in each race new needs, and, consequently, changes in their habits. As soon as one has recognised this incontestable truth, it will be easy to perceive how the new needs can have been satisfied and the new habits taken on, if one attends to these two laws of nature, which have always been corroborated by observation.

"FIRST LAW.—In every animal which has not passed the limits of its development, the more frequent and sustained use of any organ gradually strengthens that organ, develops it, increases its size, and gives it a strength proportional to the use in question; while the constant disuse of such an organ insensibly weakens and deteriorates it, progressively diminishes its faculties, and finally results in its disappearance.

"SECOND LAW.—All that nature has caused to be acquired by or lost to individuals through the influence of the circumstances to which their race has long been exposed—and therefore through the predominant use of an organ, or through the constant disuse of a part—she preserves, by reproduction, for the new individuals which come from them, provided that the acquired changes are common to the two sexes, or to those which have produced the new individuals." (Vol. i., p. 234.)

So far, except for the allusion to the "*sentiment interieur,*" the conception of need is simply that of necessity, and we have no psychological hypothesis. But the principle later develop into desire in such instances as the following :—

"The bird, which is attracted into the water by need, in order that it may find the food by which it lives, spreads out its toes when it would strike the water and move over the surface. The skin which unites the base of the toes gains the habit of stretching, because of this ceaselessly repeated spreading of the toes. Thus, in course of time, the wide membranes which unite the toes of ducks and geese are found as we see them. The same efforts to swim have spread out even the membranes between the toes of frogs, turtles, otters, and beavers. The bird, on the other hand, which is accustomed by its manner of life to sit upon trees, and which comes of individuals which had all contracted this habit, has toes which are necessarily longer than, and differently formed from those of the aquatic animals above mentioned. Its nails have become lengthened in course of time; they are sharpened and bent into hooks in order to clutch the branches upon which the animal so often rests. Even so one feels that the bird of the shore which does not care to swim and which nevertheless needs to approach the edge of the water to find its prey, is continually in danger of sinking into the mud. That bird, therefore, trying to avoid plunging its body into the water, makes every effort to stretch and to lengthen its feet. The result is that the long continued habit of stretching and lengthening its feet which is contracted by that bird and by all of its race, raises the individuals of the race, as it were, upon stilts, inasmuch as they gradually gain long and naked feet. . . . Should an animal make repeated efforts to lengthen its tongue, in order to satisfy its needs, the tongue will acquire considerable length; should it need to seize anything with that organ, the latter will divide and become forked. . . . Needs, always caused by circumstances, and the consequent sustained efforts to satisfy them, can do more than to modify organs, for they can even displace those organs when some of the needs make this necessary." (Vol. i., p. 248.)

Now, all this doubtless appears very ridiculous, and, though it is as good as any theory of transformation, so it is. But it reveals one thing, a haunting sense on the part of Lamarck that he must bring in the conception of need at every point. These are no facts which he is relating to us, they are a set of the most various and confused fancies as to how *need* can bring about the adaptations of organic life. Of the fact that need effects all this, he is well assured, but his knowledge goes no further. And he finds it extraordinarily difficult to imagine how the indispensable principle of his theory actually does its work. Sometimes that which is needed is represented as actually thought of by the animal, sometimes as merely present to its " inner feeling," and sometimes as belonging to the animal only in one respect—in that it would be well for the animal to have it, though it has it not. Sometimes the creature needs the particular structure because of other habits or structures which it has already, and which could not exist in fact without that which is represented here as derived from their need of it. In a word, the main principle of a biological system could not well be more formal and all-inclusive, or, in its working out, more indefinite.

But what is it to need? Often it is at once to have and not to have, as a creditor is without his money. Of an ill-composed picture you say that it needs a certain colour here or there, in order to complete its scheme, but not because it merely has not that colour. A parasite needs a host, but the host, although without any, does not need a parasite. Every part of my body needs the other parts. Each stage in ontogeny needs the coming stages, because they belong to its whole. On the other hand, I do not need four arms, although to have them would confer on me certain advantages. A politician needs his seat in Parliament, because he is already a politician. Sex needs its mate because they are two parts of a whole. Latent qualities are needed until they appear, and when they have appeared they are no less needed; and in both cases they are needed because they are already within the ideal unity of the organism. A cart needs *its* wheel whether the wheel is there or no, because the wheel belongs to the completeness of the structure, and fulfils the uses of the cart. In a word, A needs B because B is implied in the rest of the whole, that is in A, and because A and B are included together in our scheme. It makes no difference whether A is, to our eyes, in complete possession of B, or is completely without it.

And if an individual is without a certain quality which will yet appear at a later stage, it needs that quality in order to its completeness, but it does not need it more now, when it has it not, than later, when it is presently qualified by it. Need, therefore, is not a matter of time at all. An organism needs all its qualities and parts equally; and to say that it needs them at such and such times, or in such and such places, that is, in their proper relations and successions, is merely to say that they are such and such periods or parts; the need is the mere expression for the unity of the organism. It is even evident that this or that may be necessary to an organism but may never appear in it at all. An animal cannot be both male and female, and a foliage leaf cannot at the same time be a petal, for these forms exclude one another. But both sets of qualities are in each case necessary to the completion of the scheme. That is, qualities which need one another may be separated from one another in different organisms, or in different parts of the same organism, or in the individual and its environment, without in any way affecting the conception of need as applied to them.

The thing needed is possible to, and is implied in, the qualities of the thing which needs, and there is no question of its being present or absent in temporal fact, or as represented by a desire. In other words, need is the attitude towards one another of the parts of a real system, as soon as they are distinguished from one another, whether the distinction is made in terms of space, or of time, or in any other manner. It is the union of two differences in an underlying identity of scheme and of end, and it is only because of that reference to a common unity that need is more than a mere having or not having. Now, that unity of system is an ideal unity, and the relation of necessity is an ideal relation. Necessity, as the form of the reference of

L

qualities to one another, within an individual, has primarily no psychological reference. It does not necessarily exist as a conception of the thing needed within that which needs. Organic needs may, indeed, affect the state of feeling, and much of pain is due to the sundering of organic relations, and much of pleasure to their satisfaction. But the whole question of feeling is apart from the present subject, because feeling is not essential to need, as the latter conception is used in the theory under discussion. Not all needs associate themselves with a definite state of feeling, and an embryo is not in misery because it is not adult. Lamarck's animal needed a forked tongue, but whatever change the growing forked of the tongue might make in the emotional tone, it would hardly change it merely in the direction of pleasure. Use only hypertrophies and coordinates, but for this tongue we need some other principle than use. Undoubtedly, the "inner feeling" would come in here. Even if the being without what one needs leads to a sense of uneasiness, it does not thereby give rise to a conception of what is wanting. One may be sensible of discomfort, and yet not know what one wants; for all craving is easily misinterpreted, and may be attached to almost any object by suggestion. And science is silent as to the determination of even action by feelings without conceptions, and how much more as to the determination of growth by that means. But in any case, Lamarck's need does not always include desire; it is the mere form of organic relations.

Yet, when that form becomes a special principle among others, in the modification and transformation of species, this inevitable result occurs. Need ceases to be the universal form, in order to become a part of the content among other parts. Instead of being an expression for the unity of the organism, it gradually becomes one difference among the others, and we return to a familiar point in the anatomy of hypothesis. As such a difference, it can hardly be represented otherwise than as a psychical difference.

THE THIRD POSTULATE OF BIOLOGY. 163

Need thus becomes phenomenally operative as a conception of the thing needed, for, as an ideal relation, it cannot be quasi-phenomenally represented otherwise than as a state of a consciousness. The error is that old one of attempting to give it a place as a part at all. While, therefore, Lamarck is content to merely indicate that what is there is there as something needed, necessary, and that it subserves ends, and enters into an ideal system, he is not carrying his general metaphysic of life beyond its rightful place. But when he attempts to render this justifiable conception of organic relations into the terms of a quasi-physiological account of the origin of the related differences in time, he is merely giving us a picture-logic, which must remain for ever empty and formal, and which cannot, as an account of events which are supposed to have taken place, be either refuted or proved. The transformation of species, with him as with the others, is merely a procession of events which cannot be observed, in which the unity of the organism is incarnate as one difference among others, compelling those others into the form of organic relations.

A certain set of circumstances necessitate certain objects for which they are the special circumstances, and that necessity brings about the existence of those objects, in such and such a manner. That is the whole theory, and the special manner is, in every case, the object for the most abstract conjecture. Now, it does not seem idle to point out, as against this formula of explanation, that the circumstances are only such in relation to the objects for which they are circumstances. In short, the adapted organ is there as soon as the circumstances which necessitate it are there. The latter are only special "points" of the environment in relation to those organic parts by reference to which they are distinguished and understood. That to which it is already adapted and nothing else is a creature's environment. Animals do not first support themselves in the air, and then gain organs for doing so, and if a bird swims on the water, it is already fitted so to swim in its own manner. The organ is not needed by the species unless it is there in the species, and there to make the circumstances to which it is adapted. But to conceive of organisms as being thrown into circumstances which are not *their* circumstances, and as, consequently, standing in need of some qualification which is not theirs already, is merely to play with abstractions. There are not, in nature, such abstract organisms or such abstract environments. The matter may be made one of mere observation.

Gonzales Zarco maroons his rabbit on the island, and four centuries later its descendants are, in this and in that respect, unlike our rabbits—a matter, apparently, of new circumstances, new needs, new actions, and the whole of the Lamarckian or other formula. The rabbit, made by its needs in the old environment, is here surely remade by its needs in the island. But it is brought to England, and is found not to be a new rabbit at all, but just the same old one. The very individuals which were brought to this country returned, in four years, to the almost complete image of their ancestral form—of their form for England—for the mainland. Now, the island of Porto Santo may differ to any extent from the mainland, but the rabbit's environment (if we use the word in Lamarck's sense) evidently does not essentially differ in the one case and in the other. The animal took all its circumstances with it. Over four hundred generations among utterly new conditions could not do more to alter the species than could be undone, by a few seasons, and without generation, in this country.

Or consider Naegeli's great parallel experiment. He brought a great number of species of Alpine Hieraceae from their native elevation, in which they had lived for I do not know how many years. They differed in well marked respects from their relatives on the plains. And the question arises whether that difference was a difference in the plants as abstracted from their environment. Assuredly it was not;

THE THIRD POSTULATE OF BIOLOGY. 165

nature gives us no organisms abstracted from their environment. Naegeli planted and sowed them in the Munich garden, and, from the very beginning, they were immediately and quite indistinguishable from the forms which had always lived on the plains. The Alpine "characters" disappeared at once, and did not return. Now, we find in these Porto Santo rabbits, and in these Alpine Hieraceae, just that experience which Lamarck considers to be the basis of all transformation; yet we find no transformation. New circumstances, new needs, and a consequent change in organisation, these are to be the successive steps in the secular modification of species. But if that change in organisation were thus something separate from and only mediately following after the new conditions, then we should not, I suppose, find that these organisms immediately returned to their other form, when removed to their other country. If the organism were something separate from its circumstances, if it were only externally acted upon by the latter, then we might expect that those changes in form which, by the hypothesis, were slowly and indirectly acquired, would be equally slowly and indirectly effaced. But, in fact, the diverse conditions only affected the species for the time during which they were submitted to them, and there was no alteration of the species as existing abstractly and independently of its environment. And, in fact, species do not so exist.

In a word, the whole Lamarckian argument does nothing but to split up relations which, to research, can only be different aspects of the same forms and events, into an ingenious collection of separate things, and of temporarily separate events. Just as the form of necessity becomes a quasi-phenomenal difference of need or desire, so does the formal distinction of life and *its* circumstance become a quasi-phenomenal difference of an abstract life, and a circumstance which is not its circumstance, in an external action upon one another. And just as the Lamarckian *need* does not exist for research, so neither does the Lamarckian *environment*.

For, if a plant which has lived in warm and damp places be transplanted to a dry and cool elevation, we certainly find changes in it. But all those changes are results of the kind which biologists call direct (there are no other kinds in the world), and research works with the perfect assurance of finding this and that physiological change to be intelligibly the result of this smaller air pressure, that greater wind, that lower temperature, and so on. And if any one finds it an extraordinary thing that those direct results should be such as are suitable for the life of the organism under those circumstances (if they are not suitable the thing just dies), he may take himself to a well-worn metaphysical discussion; but there is one thing which he must not do. He must not, like the authors whom we are considering, invent and place among the direct relations which are known to research other indirect relations between abstractions, such as the species, the environment, the need, the struggle, the unity of the organism, and the adaptation, which, by their ingeniously elaborated quasi-phenomenal interaction, produce the proportioned and the advantageous in organisms.

USE-INHERITANCE.

This factor of organic evolution is not subject to most of the disadvantages which apply to the principle of natural selection, and its argument is much more subtle than is that of any other factor. For if acquired characters are inherited, then the progressive functional adaptations of the life of the individual are the variations for its offspring. These variations are therefore determined in the direction of greater adaptation to the conditions under which the species lives. Variation is therefore not so much a change as a resemblance, it is an effect, and, at least formally, an intelligible effect, of the experience of the parent.

THE THIRD POSTULATE OF BIOLOGY. 167

Over against this theory there arose the thesis that acquired characters are not and cannot be inherited, and a certain controversy which is probably familiar to the reader is still carried on, though with abated vigour, over the point. The best definition of the two forms of characters is that of Delage. He says, "Innate characters are those which have been contained in the fertilised ovum in some form or other; whether that form is known or not matters little. Acquired characters, on the other hand, are those which have been developed only through the action of the surrounding conditions." And again, "An acquired, as distinguished from an innate character, is introduced into the organism without having been present either in the ovum or in the spermatozoid; and one must also add, neither was it present in the fertilised ovum, for a character which should result from a combination of rudiments contained in the sexual elements would be innate and not acquired, although it would show itself for the first time in the products of the union."[1] I do not think that any biologist would be dissatisfied with these definitions, which, in fact, well represent the conventional distinction which was used in that most barren discussion. I have already said that the distinction answers to no difference in the world, that there is no character which is not inherited, and that there is equally none which is not also at the same time acquired, and that characters are all the same in these respects. Is this objection to the whole point of view competent?

It is plain that the theorists distinguish between two kinds of characters which are separate from one another, and exist beside one another in the growing and adult organism. On the one hand you have those which were present, "in some form or other," within the fertilised ovum. Now, these cannot have been present otherwise than as represented by agents. For if they were there merely in the sense that they were possible to the germ's

[1] Delage, "La Structure du Protoplasma," p. 198.

development, then they cannot be distinguished from the acquired characters, which were undoubtedly possible to that development, since, under certain circumstances, they occurred. And it is equally certain that the characters were not present in the germ *as* characters, otherwise they would have been the characters of the germ, and not the characters of the organism within the germ. Therefore the innate characters were in the germ as separate unknowable agents; separate, because otherwise some of them would have been there only in possibility and not in quasi-actuality; and unknowable, because they are not there as qualities, and we cannot know anything but by its qualities.

On the other hand, you have those characters which were not present in the germ, but were "introduced into the organism," or were "developed through the action of the surrounding conditions." If these characters are to lie among those which were innate, then they also must be represented by agents. For if they are nothing but modifications of the innate qualities, then their possibility and nature is entirely dependent on the latter, and they are nothing but innate qualities which are usually latent through life. Certainly they were as truly present in the germ as were any of those which, on some principle or other, are called innate. A man may grow extra muscle by exercise, but it is man's muscle, and not fish's muscle, which he grows. And not only are the acquired characters innate, in that they are possible to the germ (and that is the only innateness of which we know anything, or can at all credit), but the innate qualities are also acquired. They are, to use Delage's own definition of acquired characters, developed through the action of surrounding conditions. It would be ridiculous to deny that aspect of them. Parts are educated by parts, and the whole is educated by the quality and quantity of its food, by gravitation, air pressure, sense-impressions, environment in general—in a word, by nothing but by those very surrounding conditions which were to develop only the acquired characters. So far, the distinction seems to be only a logical distinction, since you may look on every quality either as inherited or as acquired, but it is not a valid distinction as between different classes of facts. It does not answer to phenomenal differences, though it receives, at the hand of the theorists, a quasi-phenomenal form. And the matter might well be left at this point, only we have not as yet seen what is the impulse for the distinction, in the form in which we find it.

For you might assent to what stands above, and yet urge that the conditions which bring about the acquired qualities are known, and intelligibly related to the qualities which they educate; but that the conditions which bring about the qualities which are called innate are obscure, both in their nature and in their working. In either case one has to do with a physiology. But in the former, you will say, the physiology is a plain matter of hypertrophy, hyperplasia, atrophy, and coordination, from observed and intelligible use or disuse, as well as of the direct action of the environment upon the organism. And in the case of the conditions of the education of the innate qualities, we have to do with the whole obscure physiology of the normal individual development, with the yet secret and perhaps unattainable "*Entwicklungsmechanik.*" Now, I grant that this is the case. Those qualities to which you can assign an origin in actual process, of which you know the physiological development, you call *acquired*. That is only because you see the acquirement. Those qualities, on the other hand, for which you cannot give the actual process of development, nor relate them to the external conditions of life, you call *innate*. But there is not on that account any the less process, or any the less relation to the outer world. Your distinction may be useful and safe until you suppose it to answer to any difference in nature. For the accident of our ignorance of the processes and conditions of the rise of any quality, and the accident of our carelessness and apathy to the *rationale* of ontogeny, does not affect the fact that there is a reason for the immediate existence of every quality which you call innate, and that, in virtue of that reason, the quality is, in the rise of the individual, acquired. Nothing is merely inherited; it is also worked out, and the working out is the acquirement. And if all qualities are thus, and in the same sense, acquired, so are they also, all in the same sense, and as follows, inherited. For as we earlier found that the separate agent for separate characters was an unworkable hypothesis, and incompatible with observation, we cannot allow that innate characters are in the germ otherwise than in possibility; but the characters which are called acquired are also possible, and are therefore innate.

But you may still say that the distinction in question does answer to a phenomenal difference, in this respect, that, while both acquired and innate characters are alike innate as regards their possibility to the germ, and are alike acquired as regards the actually conditioned processes of their appearance, yet the innate characters are those which are inevitable, usual, or normal, while the acquired characters are not necessary, and are exceptional and referable to distinct disturbances of the organism. To that one might answer, that the concrete organism knows nothing of the usual or of the normal; and that the mere fact of the innate qualities being common to a great number, and the acquired qualities belonging only to a few (and that only because of their special circumstances and habits), does not warrant us in the least in drawing a distinction between them as between different kinds of qualities. In the individual, a quality is both inherited and acquired, even though it were always present, or were the only example, in all organisms and in all time, of that quality. The unusual differs from the usual only very superficially.[1]

[1] *Cf.* Dr. George Wilson's paper on this subject in the *Journal of Mental Science*, October, 1896.

Everyone admits that it is probable that the experience of the parent will, in some way or other, affect the germ, and therefore the offspring. The only point of disagreement is as to the nature of these consequent changes in the offspring. Are they, or is there any reason why they should be, or could we tell if they were, identical in kind with those changes in the parent which gave rise to them? For those changes in the parent are said to be purposeful reactions to his special conditions of life, and it is argued that the offspring may, if acquired characters are transmitted, be more adapted to the special conditions at the start of life, and may need less experience of becoming individually fit for them, than was the case with its parent. Will, for instance, the child of a man who has gone to live at high altitudes, be born, on that account, with a comparatively large thorax? Or, will a pianist's child be born with comparatively well-developed digital muscles? And will this comparative or unusual quality be, in each case, the leading comparative or unusual quality in the child?

It is very hard to observe facts in this relation, and, since the manner of organic life always admits of a selection, I cannot conceive an experiment which would throw light on such a question, however long it were sustained, and however carefully it were executed. That is to say, the matter cannot be elucidated by the mere comparison of marks in parent and offspring, even for many generations, and in exquisitely determined conditions. The only method of exploring the question would be through the whole physiological history of the germ, and of its development. If you will permit me to use general terms for those of the hypotheses, I will say that as there is no particular in the germ to which the particular of the soma corresponds, the extra development of one particular in the parent is not necessarily passed on to the offspring. On the other hand, the whole germ, as a living unity with particulars of its own, will be altered in those particulars by its trophic and other physiological conditions. And it seems to me that it is absolutely necessary that we should know this intermediate germ form, and how it relates to the soma whence it comes, as well as how it relates to the soma which springs from it, before we can say what degrees and kinds of effect the particulars of the parent have on the far other pattern of the particulars of the germ, and what degrees and kinds of effect the particulars of the germ have upon the particulars of the embryo. For here is a case in which the *whether* is hopeless without the *how*, for the latter determines the possibility and the nature of an effect upon the offspring from a particular quality in the parent.

If you record sounds with a phonograph, you do not receive again the identical sounds. Some are omitted, and others are accented beyond their proportions. This coarse example is valid enough in so far as it indicates that the qualities and physiological processes of the germ must exercise, in transmission, as it were, a selection and a distortion of marks; and that this distortion must seem arbitrary and accidental to those who do no more than to compare the marks at each end of the process, and do not, in theory, even formally allow for the limitations and particular qualities of what we may call, very crudely, yet with high authority, the apparatus for transmission. It seems probable that some acquired characters may be transmitted as such, and that others may not be so transmitted, though they have their transmission in another form. We have here, in fact, but the old difficulty which arises when the organism is looked on as made up out of its characters, as put together out of separate qualities. For a minimal physiological change in an organism may give you a great and conspicuous mark, and a great and important physiological change may not appear externally under our morphological, colour, or other schemes.

The most serious difficulty with which the theory meets is, that many marks and structures do their functions without being physiologically used. Now, parts are to be explained, as regards their rise in the individual, by that physiological use, or, in other words, in terms of nutrition and stimulation. They therefore may immediately exist without any reference to their biological "function." We may justly infer from such a case, that the biological function, or external use, may be quite accidental to the organism, in that it cannot be made the ground for the present existence of the part. And if it is accidental to the rise and present existence of the part in the individual, it is not easy to see how it could at any time, or through any factor, be otherwise than accidental to it. There are two ways out of this difficulty, each of which is as fatal to the theory as the other. You may, on the one hand, relegate such parts and marks to the sphere of the rival factor, or you may suspect that we have not, in such a case, got hold of the true and concrete use of the part. But, in fact, the principle raises that difficulty for itself. For it insists on finding an ultimate physiological rationale for separate proportions and other marks, which are certainly not capable of being directly interpreted in terms of actual process. It makes shipwreck upon the self contradictory principle of the complete and self-dependent existence of isolated characters side by side with one another, whether they be physiological, morphological, or what you please.

No credible examples of the operation of this factor, even in the sphere of conjecture, have been given, except for a few simple hypertrophies and coordinations. That is because there are so few "characters" which have any intelligible derivation from their abstract use. Yet so long as this external use is held to be the beginning and end of explanation, I do not see that any better principle than that of use-inheritance is open to us. It is better than that of natural selection in the following point. Whereas the latter gives us the external adaptations of parts to one another, and to the environment, as having happened purely by accident, the former gives them to us as having been externally adapted to one another during the lives of certain ancestors, by an hypothetical set of physiological processes. Each, alike, abstracts from the rationale of ontogeny, but the principle of use-inheritance affords us a rationale somewhere.

NATURAL SELECTION.

Natural selection is the principle by which advantageous qualities lead to life and to reproduction, and the want of them to death and to elimination; so that, in a given environment, and in a given rivalry for the means of subsistence, the organisms which are best fitted in certain respects for certain functions escape destruction. This principle thus secures the perpetuation of advantageous variations. The gradual modification of species consists of the accumulation of such changes as become possible and are advantageous in relation to every change in the environment. The variations which offer themselves to natural selection may, so far as the abstract statement of the principle is concerned, be fortuitous; that is, it is not necessary that they should be intelligibly related in their origin to the functions which they are to subserve. The principle of natural selection has been so admirably expounded in many writings that it is not necessary to illustrate it in this place, nor to clear it from certain objections which appear irrelevant. But it is not to be supposed that the matter is an easy one. The principle is difficult to understand.

It is easy to make a statement of natural selection. But the principle is supposed to be nothing more than a general expression for a great number of different events in the lives of animals and plants—events which all lead in one direction, that of greater fitness to their conditions. But the manner of these facts as occurrences in nature cannot be defined or described in detail. I can fully understand the position of those who regard natural selection as the sole factor in the development of all adaptations, and in the modification of species; but I can also understand those who hold that natural selection is, first, inadequate even to the formal explanation of these things, and then a mere name for hypothetical events which never occur at all. It is important that we should not base our judgment of the principle on its undoubted ingenuity or on its facility. There may be nothing incomplete or *prima facie* self-contradictory about Darwin's world, and yet it is not on that account necessarily the real world.

Natural selection alone is sufficient to account for organic evolution. Or, it has a place along with other factors in the development of new forms of life. Or, its effect is merely to inhibit certain—and, after all, perhaps imaginary, because always inhibited—variations from the type, and to make for the stability of species. These are very different positions to hold, yet you have well-known names behind each of them, and many who unite in giving the utmost respect to the name of the principle differ even so much as regards the actual content which they read into it. And this fact alone is curiously significant in relation to the supposed derivation of the principle of natural selection from facts which have been observed.

Now, the struggle for existence is that upon which natural selection primarily depends, and it is that which we must first consider. It is a conception which at first sight includes so many and such different events in the organic world that it is not easy to find what is essential in it. We are probably safe in saying that, in the Darwinian sense, a struggle may be made out between organism and organism, species and species, and between species and environment in every case and in all respects. Even the parts of a creature, as Roux has taught, and as even Weismann has now come to believe, struggle against one another. The struggle, as between two forms, may be traced through the most complex and indirect paths. Of course, no one in this connection means an active conflict, or a direct and immediate slaying of one another. In any case, the relation receives a more or less statistical form, as indicating tendencies to interfere with one another, rather than a set and actual interference among their activities and changes. On the whole, organisms do not seem to do this and that, and also struggle with one another, but they struggle with one another in doing this and that and all the rest.

In a word, whatever they do, organisms and the part of organisms struggle with one another and with their environment. The conflict is an aspect of all their qualities and activities which have an external use. Find a relation, one may say, between species, or between individuals, and you will find that this relation, merely formally and as a relation, may receive the form of the Darwinian struggle. By swiftness and cunning, by strength and weapons, by beauty and valour, as well as by ugliness and cowardice, by every instinct and habit, and by every form and quality and part that can be named, creatures struggle for existence. Against cold and heat, damp and dry, and all natural features of the earth, as with their environment, all organisms continually struggle. The struggle for existence is one form for all organic relations, and there is no quality of organisms which, by acuteness of research, or by ingenuity of conjecture, may not be found to have a certain use, and therefore to be a certain weapon. The very relations of members are, as we have seen, not incapable of this very form. The germinal layers, and later the tissues and organs of the body, and indeed any elements into which you may, on any principle whatever, divide it, struggle amongst themselves for food, for space, and for representation in feeling, so that this struggle has been made into one of the factors of ontogeny, as an event which takes place. The mother struggles with her child for nourishment. All individuals of one sex struggle with one another for those of the other sex. Parents struggle for representation in their offspring, and even forgotten ancestors, we are now told, are separately within us, conflicting among themselves for another sight of the sun. Latent and alternative qualities, as we saw in the last chapter, struggle together for appearance.

You will say that we have left the struggle of the natural selection doctrine. I think not; but in any case we must go further before we come back. For the conception of just such a struggle (I do not say selection) is not confined to biology. Ideas are represented as existing separately, mobbing round the trap-door under the stage. Motives struggle together, and so do faculties. Tendencies never do anything else, and motion in a curve is the conflict of two movements with different accelerations at right angles to one another.

Now, in all these cases—take the form which is common to them—the struggle is *in no case* what happens or what we see. What happens and what we see is in all cases the result of the struggle, and the latter is a certain form of explanation, quite just in its way, but unable to boast of observation. The struggling elements are the ideal abstracted components of the event, and the method in general is nothing but the finding of the imaginary straight coordinates of the curve. And this is the case no less in the Darwinian struggle between a species and its environment, than in the conflict for nourishment between a mother and child. It is not from observation that we find, in the latter, a pair of hungry wolves, but because we know, on some principle, that they must be so, just as we know that the members of our bodies must struggle together for space, for stimulation, and for nourishment, and just as we know that a species must struggle with all other species with which it comes in contact. Thus we may take together all forms of the struggle, the most peaceful with the most turbulent, inasmuch as these characteristics are merely accidental to the meaning of the conflict, as it is used in theory, and as it does not take place, as a set of special events, in nature.

It comes to be the question, why should we speak of two things as *interfering with* one another, rather than as being *related to*, or conditioning one another in such and such a way?

You justly say of a honeycomb pattern that its six-sided figures are circles pressed together. Only, you do not mean that they are in fact circles, or circles otherwise than in your imagination. You mean that the figures might be ideally constructed by conceiving separate circles to struggle together for space, until they should come to the honeycomb equilibrium. In the case of the many-sided bubbles in a frothy lather, you may, more truly, speak of spherical elements struggling with one another, for, if one were taken alone, it would of course be spherical. Only, in fact, no one of them is alone, or is, otherwise than in your imagination, spherical. And if you say that there is not room enough for all the bubbles, and that they are spherical, I shall answer that the bubbles are many-sided, and that there is room enough for what bubbles there are. Now, you, for the moment, are looking at the bubbles as unrelated to one another, as ideally existing separately from one another, and as, therefore, externally conflicting with one another; but I, for the moment, am looking at what is in fact, and am taking the bubbles as they concretely exist, and are made by one another. For you say that they struggle against their limits, but I know them only by their limits, and, indeed, find them to be nothing but their limits and conditions.

You justly say that the mother and child are continually struggling with one another as two separate individuals who draw upon a common supply of food. But you only say this so long as you look on them as two separate organisms, not essentially related to one another, yet both requiring support from a third abstraction, the common fund of nourishment. Research shows you nothing like a struggle, because research, as we have often seen, knows nothing of such isolated and abstract agents. These two are not more truly two than they are one, and there is no more reason for speaking of the division of the fund of nutrition between them than there is for speaking of the struggle for food between the organs of one body. Yet even these organs are looked on as abstract unities, and as, in consequence, only externally interfering with one another, that is, struggling together in those respects in which they are related. It is only by denying the actual relations between mother and child, or between organ and organ, that one can come within sight of the point of view of struggle from which these differences in unity appear as mere differences in contingent collision. What happens and what we see is an intelligible harmony. But it is open to us to regard certain elements in this intelligible relation as limited, cramped, and repressed by a struggle with the other elements, just because there *is* a relation, which is thereby formally regarded as unintelligible. And that point of view is arrived at by looking at the elements, not as they are, for they are nothing but their limits, but as ideally existing without limits or relations, and as externally and accidentally encroached upon.

You may truly say that, in the embryo, the alternative characteristics of sex struggle together for the privilege of becoming phenomenal, and you can, as we have seen, quote in your support the masters of biology. But the condition of your doing so was found to be that you should postulate a real (quasi-phenomenal) existence, as for a separate object, for something which no more really exists than does the sphere in the bubble of the foam. You must, if a sex-character is to struggle and conquer or be repressed, regard its ideal presence in the whole scheme as the presence of an agent which is its inner nature, and is itself absolute and unconditioned, except by the external struggle. If the male qualities are to struggle with the female until one set appears, then we have to do with an hypothetical process which we do not observe, but which we judge to exist because of what we call its results. And the nature of our judgment is that we mistake ideal abstraction for physiological analysis, and postulate a real struggle which can never exist because its elements can never exist.

And it is in no other way that you speak of the struggle of species with species, or of species with the natural features of the earth. The tendency of reproduction to outrun the means of subsistence is one side of the war between a thing which is regarded as without relations or limits, and those conditions themselves. The latter, therefore, appear to be mere cramps and fetters externally imposed. The struggle for existence is the imaginary struggle of those movements into which rest may be resolved, and as the latter struggle does not occur and is not seen, so neither does the former occur or appear. The view of nature which regards organic differences as in continual direct or indirect conflict with one another is, as regards nature, quite gratuitous, and, as regards science, quite abstract. A species is at no time, in fact, more numerous than can be supported by its means of subsistence, and it seems probable that it never comes near to such a limit. "But it has enemies," you say, "and other limits to its numbers; it has a low rate of increase, or the young perish in great numbers; and so on." Certainly that is the case, and no one would ever deny that the numerical strength of a species is intelligible in relation to the conditions of its individuals. The individuals are related in infinite ways and directions, and as the conditions change, so will the number of individuals change. But it is only so long as the inquiry remains purely formal and hypothetical, regarding the species, falsely, as an individual or agent, or, becoming a research, remains merely on the abstract statistical plane, that the appearance of the Darwinian struggle can be left to us. Abstractly, one can look on species as bodies which are always trying to expand, but which come into collision with one another, and so remain at an equilibrium of tension. But that "trying to expand" is, as an account of fact, merely imaginary. What is an attempt or a tendency which is not a fact and an event? Now, this attempt is always inhibited, and it never becomes a fact. Nor does the supposed complementary repression of it become a fact. The struggle between species or between the members of a species, being, as we understand, a conflict by means of all qualities which have external uses, is no more a special phenomenon of natural history than the struggle between the members of my body is a fact of physiology. In either case we have to do with nothing more than with a merely general anthropomorphic expression for relation.

The conception is formal, not by the accident of our ignorance, but of necessity, because there is no relation of individuals which does not come under it as fully and as well as all the others. In so far, for instance, as an organism is adapted in this or in that respect to one set of conditions, it is probably ill adapted in this or that respect to another set of conditions. Thus, in every respect in which the creature may be said to have an environment, it is locally and otherwise confined by the limits of various circumstances. Specialisation of any kind imposes answering limitations on the species, and it is against these limitations that the species is supposed to struggle. But these limitations belong to its character; they are, as it were, its obverse; and, unless we are to represent an organism as divided against itself, we cannot represent it as divided against its environment. For we may put the whole of the Darwinian struggle into this form, "an organic species strives to expand numerically, and the environment of every kind strives to repress its numbers; that is to say, an organic form in all its relations strives to expand numerically against those limits which belong to and may be said to be its form." Now, this strife, being universal in organic relations, is not a particular relation here and there. And, in order to arrive at this expression of all the relations in the form of a struggle, you have only to follow the familiar Malthusian method of abstracting reproduction from its conditions (which are itself), and then bringing those conditions externally to bear upon the supposed unconditioned reproduction.

The struggle is thus not only a universal form, but it is also one which cannot be filled, and is therefore inapt for research. For, to fill in the form destroys it, and to give a detailed account of a relation prevents you from any longer calling it a struggle. For the struggle is between abstract agents without conditions. And research, as we have seen, destroys all hypotheses which work with such agents, simply because research establishes concrete relations at every point. Thus it destroyed the mere difference or discrete quality of the first postulate, and the mere identity or anthropomorphic agent of the second. And thus, in the third place, it knows nothing of the elements of the biological struggle, the unconditioned species, the abstract organism, or the complementary abstract environment. For all these are fictions; they are metaphysical distinctions which have received, at the hands of the theorists, the form of quasi-phenomenal difference.

How little the elements in these struggles really exist, and how little the Darwinian struggle is itself a fact of natural history, may be seen from the words of that biologist himself. After remarking that there must be conditions which check the increase of a species, and sometimes bring it to an end, he goes on to say: "If asked how this is, one immediately replies that it is determined by some slight difference in climate, food, or the number of the enemies; yet how rarely, if ever, we can point out the precise cause or manner of the check." In other words, one immediately replies with the form of the struggle, yet how rarely, *if ever*, we can fill up the form in the least, or point to any concrete struggle at all. As I said, the struggle for existence is nothing which exists or which we see.

It has often been urged that this conflict does not signify any actual enmity, dissension, privation, or any struggle in the sense of direct physical opposition, and that nature hides her frowning providence under a smiling face. And it is only just that we should remember this, its author's own caution, in the study of the theory. But we may also notice that if the relation of individuals or species is to be called "interference," rather than "condition," or even "cooperation," there ought to be some sanction for the special term. Now, that sanction could hardly be given except from observation of just such natural atrocities as we are told not to expect and to leave out of account. Or, at the very least, it requires that one should be able, more certainly and more constantly than is implied in Darwin's "rarely, if ever," to point out the "precise cause and manner of the check." We might yet be persuaded by a tale of natural atrocities and catastrophies. But such a tale does not come before us, and Dr. Stirling's study of the struggle, which he bases wholly upon the comparatively insignificant part which is played by slaughter in nature, seems, therefore, to be relevant and complete. This struggle, he shows, does not correspond to any observed events and processes in nature, whatever else it may correspond to. After exhibiting Darwin's impressions at the time of the journal, Dr. Stirling well says : " So far, we have, on the part of Mr. Darwin, one sole reference to life, life infinite in its numbers, infinite in its varieties, and there is not as yet a note, a hint, a whisper, of those mortal straits in bitter struggle from whose fatal pressure only the fittest emerge. No doubt there is strife-life in some only through death in others. But yet scarlet —blood—cannot be called the colour of the scene. There is infinitely more of a smile in it than of a shriek. What is savage is, in its paucity, out of all proportion to what is tame." (" Darwinism," part ii., chap. v.)

Delage, in his remarkable work of 1895,[1] sets down in order those objections to the commonly received views as to the adequacy of natural selection, which have seemed to him and to others to be insuperable. These objections combine, according to him, to prove that the principle cannot give rise to new species, and that it is not adequate to the explanation of organic evolution. But, as they are fully as conjectural as Darwin's principle itself, they do not seem to be very final. They stand over against the irrefutable proofs, and never come to mix with them. Their value seems rather to be, that they show that one may uphold almost any point of view upon this plane of conjecture. The author, in introducing them, remarks that "natural selection is an admirable and a perfectly just principle—a point upon which the whole world is agreed; but the limits of its power, and the possibility of the formation of new specific forms through its agency, are still matters of debate. It seems, indeed, to have been thoroughly proved that the principle has no such power."

Yet, for just that task and for no other, was natural selection introduced into the language. When its occupation is gone, one wonders in what sense the principle remains "admirable and perfectly just." All that one can say about it, except that it merely occurs, is here denied; but its unqualified occurrence is not denied. And this is so because you cannot deny a form like that of the struggle for existence, as you can deny a fact. The strength of the Darwinian theory is that it tells us nothing; not that it tells us things which can either be proved or refuted. The objections to the principle, most of which have become classical, are as follows:—

"1. The causes of variation are weaker than those of fixity; and the latter must therefore overcome the former.

"2. Selection is impotent, because the greater number of the characters which it is supposed to have developed are useless, and thus give no opportunity for its action.

[1] "La Structure du Protoplasma," p. 370.

"3. There are numerous useful characters which selection cannot have formed, because their usefulness only shows itself when they are fully developed.

"4. Variations, even when they are useful at every degree, are not useful enough to create an advantage which can be seized by natural selection.

"5. The selection of fortuitous variations cannot give rise to species, because these variations are isolated, and because, in order to give rise to a real advantage, they would have to bear on several characters at one time.

"6. Selection is impotent because the variations upon which it might take hold are ceaselessly destroyed by sexual intermingling.

"7. Selection is not the true cause of the formation of species, because, if it were real, however feeble were its effects, it would transform a species in much less time than that which is evidently necessary for such a transformation; and, for the transformation of a species in a reasonably long time, the necessary protection is so weak that it becomes illusory."

Certain of these objections are little more than a criticism of the ordinary use of the "character" as the ultimate unit of biology, and they all show the impossibility of arriving at any agreement upon these debated "factors," except by means of a general theory of the nature of the qualities of organisms. Take, for instance, the second of the above objections. What are these useless characters which selection is supposed to have developed? Clearly, the generic morphological characters. Delage quotes Naegeli as saying that "if the theory of selection were true, the most useful characters ought to be the most firmly fixed. But quite the opposite is the case. The most constant characters have always an anatomical arrangement, independent of adaptation and utility. Such are the arrangements of leaves, opposite in the Labiatæ, spiral in the Borragineæ, the division of the terminal cell, in transverse planes in most Algæ, in oblique planes in certain mosses and in the vascular cryptogams, and so on. Darwin admits that these are indifferent characters, which have been fixed, at one time or another, by the nature of the organism and the surrounding conditions, and have become constant without the aid of selection. But," says Delage, " if such characters can have been fixed in such a manner, the others may have had a like origin, and selection becomes useless."

Here we have no distinction made between morphological and other characters; they all lie side by side as similar units. If such speculation ever thought of natural conditions and processes, it would not fail to see that the proportions of an organism, for instance, are of a very different kind from its colour. Further, we have, in this second objection, the astonishing premiss that the greater number of the isolated characters are (mediately, one must suppose, as well as immediately) useless, and exist thus beside other characters which are useful. We have already had occasion to see how little, even from the biologist's point of view, this "useless" means, but we may notice that it is here thrown down as though it were an absolute distinction between kinds of characters.

But, taking these objections in order, we may find the first to be credible in point of fact, inasmuch as species do not, so far as we know, change, but baseless in point of argument. Delage shows that "if, in the first generation, one in a thousand of the individuals is affected by a particular variation, in the next generation there will only be one in a million so affected, the proportion being reduced with extreme rapidity, and soon arriving practically at zero." On what basis this calculation is made I am quite unable to see.

Of the second we may say that it is impossible for us to deny a certain use of function to any part of the organism; first, because our knowledge is rarely complete enough to do so, and then, more surely, because the part or mark in question arose and is maintained with and by the rest, and is therefore implied in the development of marks and parts which are known to have a certain use.

The third, that structures which are supposed to have been educated by selection, would have been useless in their earlier and less developed stages, is also quite as much an objection to the principle of use-inheritance. For if they were useless, then they would be unused. It is a common error to suppose that the latter factor has a wider scope than the principle of natural selection. Now, this objection shows how impossible is a science which is based on the external uses of organic differences. It further shows that no principle should be used in the explanation of phylogeny which is not also able to take its place in the explanation of ontogeny. But that is to say the same thing over again. For such useless earlier stages occur also in ontogeny. They have their function in the embryo, and the Darwinian might well urge that since this is so they may also have had their function in the adult of long ago. And in fact this is the ordinary answer to this objection. The structures are useless, in the earlier stages, for that end which they now subserve, but they were useful for other purposes of which we know little or nothing. And so the matter remains one of empty conjecture.

The fourth is based upon the Darwinian form of gradual modification. That "gradual" or "by infinitesimal steps," is a trick to cheat our perception of the profound difference of first and last. But it becomes a treacherous instrument when the selection of such minute variations is pictured in detail. For such a very small modification cannot have its special advantage in the struggle for existence. In his comments upon this objection, Delage brings forward points of great interest, but the whole discussion has its deepest interest on account of that utter abstractness which is essential to considerations of natural selection. He thus refers to Naegeli's argument: "Let us suppose that the giraffe has taken a thousand generations to gain the length of its neck. If this neck is one metre longer than that of its regularly proportioned ancestor, the gain has been one millimetre in every generation. Now, a neck which is one millimetre longer gives no advantage, even in times of famine, and even in the case of animals which browse on the leaves of trees. Darwin speaks of a variation of two inches. But, even if we admit of this exaggerated variation, one does not reach the end, for, in times of famine, the animals do not die; they suffer and grow thinner, and then fatten up when plenty returns. If some die, they are the diseased and the aged, and those which gain some advantage from any particular peculiarity are not the only ones which come out of the trouble. But Naegeli," says our author, "is mistaken, for not the diseased or the aged would die, but the young which have only just been weaned." For my part, I cannot guess which of the unfortunate animals would succumb.

Such difficulty, such absurdity, arise even in the discussion of the most simple possible illustration of the Darwinian theory. For you have here a case in which it is just conceivable that three inches more of neck might save the creature. But such cases are extremely rare, even in the works which have been written in support of the theory. "Some slight advantage" is all that is usually to be gained —not life. If the principle cannot be certainly applied to this crucial instance of the giraffe, I do not think that we can look for much help from it anywhere else.

The fifth objection has its power, because the theory of natural selection is based upon the doctrine of the inner independence of organic qualities, but it has its weakness, because it still, unnecessarily, remains conjectural. "Accidental variations cannot give rise to species, because these variations are isolated, and because, to give a real advantage, they would need to have effect on several characters at the same time." Now, however clearly this point is developed, and, however "many" variations are

THE THIRD POSTULATE OF BIOLOGY. 189

proved necessary in complement to a primary variation, we yet do not, by it, prove anything against natural selection, except by the vaguest calculation of probabilities in cases of which we know nothing. If variations are still essentially independent, isolated, and numerable, it may be difficult to imagine how the necessary assemblage of variations, in each case, comes about, but, though it is mathematically improbable, yet it is still possible ("easy to imagine") that when a given change arises, those which are necessary to it may arise within a few generations of the first, and that these, all together, may give rise to an advantage which should be considerable and efficient in the struggle for existence. And so both the objection and the answer to it remain entirely conjectural. That is only to be expected, since they alike deal with abstractions which answer to nothing in nature. And it is impossible that, in such cases, any amount of debate and any number of illustrations, should bring us at all nearer to a conclusion. But there are no "isolated" variations, as there are no "isolated" characters, and as we have already seen, the postulate which underlies all the strength and all the weakness of the Darwinian theory is an incredible thing. For that theory is there in order to explain the external adaptation of part to part. And here we find the objection that unless part be *already* adapted to part, its mechanism of hypothetical processes will not work. In fact, there is, physiologically, and as regards inheritance, no variation but the variation of the whole; and research, in fact, both expects and reveals an internal adaptation of all the parts.

I do not know whether any advantage is to be gained by the consideration of very complex and astonishing adaptations, but I will venture to say that no one will work seriously through the histology, and embryology, and physiology of any organ, and still remain satisfied with the formal explanation of natural selection. For if we must *count* as the biologists count, then the *numbers* of necessary complementary parts, and variations, and stages, in the development of the individual, in order to one another and to the whole organ, might be enumerated for ever. Natural selection is based on the analogy of artificial selection. Now, the latter seizes upon external points and insists upon them, but it never professes to externally adapt point to point. The internal adaptation is maintained in spite of the breeding, and is the limit for its transforming power. Thus the artificial selection differs *toto coelo* from the natural selection, which is supposed to effect the external adaptation to one another of parts which are internally separate and absolute.

On a page of Dr. Stirling's book there are the following expressions with regard to the eye :—" The formation of the eye by such variations as those of which a cattle-breeder avails himself!" (Lyell.) "What seems to me the weakest point in the (Darwin's) book is the attempt to account for the formation of organs, the making of eyes, etc., by natural selection." (Asa Gray.) "About the weak points I agree: the eye to this day gives me a cold shudder; but when I think of the fine known gradations, my reason tells me that I ought to conquer the cold shudder." (Darwin.) "What we have so often seen, namely, a simple casual variation as such may somehow chance to hit, quite naturally, into the conditions of its environment in some new way which shall give it an advantage in the supposed struggle for existence. *And it is in this way that an eye is created !* " (Stirling.)

Why do these authors find a special difficulty with regard to the eye? For the perfection of the eye must be of the most primary importance in the struggle for existence. Everything in it has, most obviously, its own external use. If adaptation is anywhere, it is here. Surely the eye should be the last thing which should present any difficulty to the Darwinian theory. But, in fact, here is the very source of the trouble. They know too much about the eye. The thing does not essentially differ from other organs; but in this case our knowledge of the concrete necessities of proportion and of coordination is very full. So fine is the adjustment of part to part, so exactly defined is the necessary form and histological structure and innervation of every part, that the independent variations of parts, as the material for selection, is incredible. So close are the limits to possible changes which are yet compatible with function, that those who find it most easy to imagine cannot here imagine the separate and external adaptation of part to part by fortuitous variations and natural selection. Every part does not, in this case at least, exist because of its separate significance to the individual, and merely so exist. The mere selection of this and that particular in the eye, because of some advantage which it should confer upon the species, is, as an account of process which has taken place, here laid bare in all its emptiness. It is not possible to suppose an advantage unless all necessary changes took place at the same time, and this would be no less than the alteration of the whole organism. The outer use is not attained in independence of the immediate inner relations of ontogeny and of maintenance, as the breeder's whim is so attained. The whole sum of the relations, of every kind, are one; and while every relation which you find may receive its teleological expression—so that the number of necessary adaptations is infinite—no one of these relations, in dogmatic abstraction from the others, can be looked on as the complete rationale for the part. Reason does not select things which have been unreasonably made, in accordance with their adventitious uses, and thereafter retain them, unreasonably stored in the anthropomorphic agent of the hypotheses. Much rather, it makes and informs them, and the uses of things are not separate from their immediate origins. And I do not say this merely because "it must be so," nor because of the never-ceasing recurrence, and the fascination of vitalistic or quasi-psychical theories of organic life; but rather because research ever tends more and more to find that the purposefulness of the processes of formation and maintenance and reaction is not separate from, or accidental to, their necessary reciprocal relations. Research knows less and less of the distinction between cause and end, and the problem of the union of mechanical necessity and adaptedness is one which, while it is ever present to biological speculation, recedes before detailed physiological investigation. Only, as I said, in the case of the development of the individual is such detailed knowledge wanting, and perhaps never to be had. But we are not on that account warranted in assuming that the formation of the body is not intelligible in the same way as the hypertrophy and hyperplasia of organs, in relation to a new demand upon them, are intelligible. And we are, therefore, not warranted in choosing a certain "use," for which a certain particular is made, and in referring its origin, in abstraction from all real process, to the agent immediately, and to the factors of organic evolution mediately, and in taking it as existing merely because of that use. For this is the very suicide of science. It involves, for instance, that independence of complementary variations which we have discussed.

Of the sixth objection one can say nothing with certainty. Does sexual intermingling favour variations or obliterate them? Opposite views have been held upon this point, and the union has obviously sometimes the one effect and sometimes the other. But it seems certain that while the union of different individuals may give rise, in this case, to one, and in that case, to another variation, yet if you take the whole species together, the consequence of free-breeding would be that any particular variation would tend to be eliminated by that very process which gave rise to it. But then breeding is not free, and is variously restricted in different species. Again, it is, I suppose, probable that, if a variation arose in this case, it would be likely to arise in others of the same species. But I do not see that we are in a position to know anything about the matter.

According to the seventh objection, selection is not the true cause of the formation of species, because, if it were real, however feeble were its effects, it would transform a species in much less time than that which is evidently necessary for such a transformation. Now, this objection seems to be relevant enough, inasmuch as species do not at present get transformed, in spite of the fact that independent variations still occur, and the struggle for existence continues its deadly selection. This thesis, like so many of its kind, is based upon a mathematical calculation, in which there appear "single variations," and though the calculation is correct enough in form, it must be evident that our knowledge of conditions is too meagre to serve as a basis for such an abstract method, even if the imaginary unit had any existence in nature.

Now, large volumes might be filled with the barest account of all the objections to, and the supports for, this principle of natural selection. Hartmann's "Ergænzungsband" contains a great number. Virchow, Eimer, and many other authors have reasoned admirably against it. Mivart has a genius for insuperable objections, and so has Roux. But the principle is hydra-headed; it appears in biology and in other subjects quite unaffected by the insuperable objections. It seems to be entirely a waste of energy to attempt to disprove a form which has confessedly no content, by leading against it the evidence from observation. On its plane, as upon the plane of all the biology of hypothesis, it is possible to plausibly present almost any conclusion. It is not, therefore, a matter for research, or it would have received a fixed authority long ago. It does not depend upon, and it does not assist research. But it is a metaphysical theory of organic adaptation.

And all that it says is this. Such and such (unconditioned) qualities or structures either fit their conditions or do not do so. Those which do so are those which we see and which exist, and we necessarily interpret them in relation to their conditions. Those which do not do so neither exist nor are seen, and they are, in a word, purely imaginary.

And the achievement of the method is not to explain anything which is, but it is merely to afford us a transition from the really unintelligible of accidental production, to the formally intelligible of conditioned existence. It enables one to follow, hypothetically, the production of the parts of the system of the body, or of the organism and environment, as unconditioned by the other parts of those systems. Then, at a certain point which cannot be shown as phenomenal, these parts come into collision with their conditions, and those only which fit the latter (that is, all those which exist) come to be selected. Thus the Darwinian thunder-clap follows upon its proper blaze of abstraction. And all that it succeeds in doing is to offer to us an empty formula of explanation which enables us to explain the parts as essentially unrelated to one another. In this respect, and in harmony with the first postulate, it is an alogical principle, and is necessarily, as in fact, without interest to research.

The factors of organic evolution are thus attempts to put together again the unity of the organism. For that was first, as we saw, disrupted. The second movement was to put it together again by means of the agent. This stage did not complete the whole matter, because the action of the agent was of necessity arbitrary and unintelligible, and its products were not at the same time, and immediately, the products of one another. So we have recourse, in the third postulate, to the fiction of the becoming related to one another of parts which were inwardly dissociated from one another, or, if associated, then only joined by the unexplored mediation of the idioplasm, physiological units, or immanent soul. And this is the case, as we have seen,

THE THIRD POSTULATE OF BIOLOGY. 195

not less in the principle of use-inheritance than in that of natural selection.

For when we regard the parts as existing solely by reason of their external significance to the individual, in abstraction from the actual processes of their rise and maintenance, we are reduced to these factors or to the popular doctrine of design ; and our error in each case is the same. For we are dividing into two separate kinds of events what are, to research, two sides of every organic event. The abstract necessity of the event is, in each case, given to us in the present, and the secret of its intelligible relations to the rest of the individual is put away into some distant age, it matters very little when. But if the harmony of system is due, in any way, to the external putting together of ingeniously adapted parts, which have no essential relation to one another, and are not each implied in all the rest, if, that is, each part has just this one significance which we call its function, and is no more than it—then I would rather subscribe to the doctrine of creation, as the Darwinians expound it, as the least likely to interfere with and to confuse the issues of research.

CHAPTER V.

THE UNITY OF THE ORGANISM.

OUR discussion of biological theory has centred round one point. It was not from choice that we criticised every theory by a study of its treatment of the unity of the organism, but it was because that attitude of the hypotheses was evidently the most important, and often the sole, impulse for them. The first postulate gave us the unity of the quality instead of the unity of the organism, denying the latter as it is known to research; the second was forced to reintroduce the unity as the anthropomorphic agent, or maker of unity, and the third postulate derived every separate and independent quality from its external significance to the unity of the organism, but denied the actual relations of the parts within the unity which we use in observation. A central conception such as this deserves attention. This unity, whatever else it is, is the very category of biology, and we have seen, from the formality and emptiness of its discussion, that we have to do with nothing but the discussion of a category. The unity of the organism is the impulse for the doctrine of type, and for the theory of vitalism. It is the beginning and end of the doctrine of design. And we have followed it, as fully as their interest seemed to warrant, through the hypotheses as to the determination of the form of the individual, and as to the origin of organic adaptation by the transformation of species.

I need not, therefore, apologise for inquiring a little further as to what the unity of the organism is, and what it is not. And, in the first place, it is not *any one* of the particulars of the organism. That is, it is not an agent. It is not, for instance, abstract protoplasm. There are many, I do not know how many, species of animals and plants. Now, it is commonly assumed that there is a different protoplasm for each one, and that their differences from one another might be given, with a fuller knowledge, in the definite differences of these protoplasms from one another. But consider the vast number of species, and that within each species there are innumerable individual forms which definitely differ from one another. There are also innumerable different forms which have been, and which are not ; and artificial selection shows us that there are innumerable forms which are possible and which yet are not. There are, further, yet other possible forms which would not be viable, but which might pass through the earlier stages of ontogeny, as every pathologist knows. Consider the number of these germs, and set down any estimate for it. The germ, in each case, must have, according to the theory, some difference in its protoplasm which contains the determination of the individual form, and which determines it definitely and rigorously. Then consider the number of cytologically different somatic cells which belong to these bodies, and lastly, consider the various kinds of protoplasm in each of these cells. It is an inconceivable number of different and definitely different protoplasms.

Then, on the other hand, consider the minuteness of these cells, and that these protoplasms, while they are theoretically different, are known to be similar within narrow limits. They must all, for instance, preserve the functions of assimilation, reproduction, and irritability. And lastly, that the substance which we name protoplasm is unstable in the highest degree, is always rising and falling in the rhythm of metabolism, and is most easily susceptible to outer influences and subject to degenerations of various kinds. The observed changes which take place within protoplasm during division can hardly be dissociated from chemical change.

Thus we have, on the one hand, a vast number of very definite images, which are in some way determined as to their form in the germ, and in every part of the soma through later stages. And we have, on the other hand, this substance, protoplasm, the material which these germs and cells contain, and are asked to believe that it, in all its change and instability, is the guardian of the form of the individual, and that, notwithstanding its necessary similarities in all animals and plants, as well as its minute quantity in every germ, its definite differences are the secret of the definite differences of the organic images. Even were protoplasm to remain the same, in a cell, without constant changes, constant metabolism, and constant liability to accidental alteration from without, we could still not find that there is room enough in organic germs to allow of definite difference in each germ, answering to the definite difference of its image. But protoplasm does not remain the same for an hour, and, if the common contention were a true one, species would not remain constant for a day.

I am averse to conjectural calculations, but the reader may amuse himself with this one. Set down as many millions as you like, for the different individual forms of animals and plants, and you will find that, even allowing for no similarities between germ protoplasms, and for no limits to their difference, it will be hard for you to invent a form for the germ difference which shall allow of so many definite differences in the same number of minute, and, after all, plastic germs. Further, it must be remembered that a germ, or a Protozoon, may be divided and yet live and retain intact the image which is entrusted to it, and all the evidence which we have goes to show that the protoplasm of the germ is on the whole homogeneous as regards the determination of the form of the individual. I do not think that research will ever show, or that hypothesis is justified in postulating, a definite difference or system of differences within a germ, such as is required by the belief

THE UNITY OF THE ORGANISM. 199

that the form of the individual is determined, in its rise and maintenance, by the special character of the germinal and somatic protoplasms. If the germ were not comparatively homogeneous as regards the determination of form, the matter would be somewhat, but not greatly, more credible. But, as it is, we are to imagine millions of different homogeneous characters of protoplasm, so widely dissimilar that the continual physiological and accidental changes in each do not affect its constant identity of character, still more widely dissimilar because there is in each case only a minute mass of the substance to exhibit it, and yet so closely similar that all these millions of different protoplasms are still protoplasm, assimilating, excreting, and growing. There is small wonder that recourse has so often been had to an expression of the difference of germs in the form of a quasi-psychical principle. For there is no structural or chemical difference which can be adequate to the multitude and to the definiteness of these differences of determination. Yet, as we have seen, the immanent soul is no credible way out of the difficulty.

Further, the matter is made no simpler by the fact that the differences between protoplasms which we do know cannot answer to the differences in their determinations of form. We have here only a special case of the old disagreement between morphological and physiological facts in the same organism. For as the adult morphological proportions are not fully intelligible in terms of their functions, so that we are led to distinguish between homology and analogy, so also the potential morphological proportions are not, it would seem, a function of the physiological peculiarities of the germ which is to give rise to them. For we may recognise the widest differences in the chemical properties of the protoplasms of two *nearly allied* Protozoa, and we may further distinguish between them by wide differences as regards their consistency. With the same morphological form, for instance, there may appear the widest differences as to the character of the external covering which is excreted or built up. And in the higher plants we may recognise the widest differences between the chemical properties of the seeds, and between the chemical products of the parts of the grown plant, even when the plants are morphologically closely allied.

I have no theory whatsoever as to the nature of the essential differences of germs. Only, we cannot find it in their chemical or structural or physiological properties. When a somatic cell, for instance, reproduces the whole form, we cannot fail to recognise that it does so, as it were, in spite of its particular functions and its particular structures. And we are not, on the other hand, surprised to learn that what can be known about the particulars of germs leads us no nearer to an understanding of their form-determining properties. We can only say that we do not know the essential differences of germs and of protoplasms in this respect, and that what differences we do know appear to be wholly accidental and irrelevant to this unknown essential difference. This seems to be equivalent to the admission that there is no difference of germs or of protoplasms in terms of which the possibilities of form might be, even formally, expressed; and for my part, taking all the evidence together, I believe that this is the case. Probably this difficulty has been the chief impulse for the doctrine of isolated anthropomorphic agents. For this essential difference of germ is the universal which we cannot find as a particular apart from the others. Certainly, all the differences of the germs must have their place in the determination of the forms which come out of them. Our uneasiness arises from the fact that we cannot find, under any one particular form, differences between germs which should answer intelligibly to the differences between images. And all that we can know about protoplasm is its fine structure and its chemistry. It is not astonishing that we should be able to find in either of these respects a difference which should be the essential difference in question.

An interesting question does, however, arise in relation to the stability of species, and to the definiteness of the form which is "determined" by the germ, in that the substances which we group under the term protoplasm are so unstable as we know them to be. The consideration of this fact not only leads us to lay less emphasis than is usually done upon the germ as the adequate determinant, but it has also led some to doubt whether we are not in the end compelled to introduce some hypothetical agent into the science, *on the understanding that it is not phenomenal*.[1] One may say, that it is improbable that the interpretation of form can ever be made in relation to the germ from which it arises. For instance, we have already seen that the outer adaptation of the alternative forms of sex, and of the successive forms of a cycle of generations, is so complete and is carried out with such apparent disregard of the possibility of a common homogeneous origin, that we may almost despair of finding that male and female, in all their qualities, are necessary results and the only possible results of the development of a germ which is, as regards them, truly identical. The choice between these so diverse, so definite, and so complex forms, is made by a simple alternative condition in the embryo. And we saw that if these two forms are the *necessary* alternative results of the development of one germ, then it is almost incredible that they should be so nicely made *for* one another. Of the adaptation we are sure. But our belief that the end or ends of the development are intelligibly determined by the qualities of the germ, and are so determined in every particular, is, in the end, incompatible with our recognition of the plain fact of the external adaptation, although it depends upon a metaphysical assurance which one would not like to question. The latter assumption is, that at any stage of the

[1] As in the "Philosophy of the Unconscious."

individual, germ, embryo, or adult, there is nothing present —nothing that can determine—but the qualities of that stage as they can be known to observation. It assumes that the unity of the organism is not present as one factor among others. But you see how inevitably, although I believe viciously, one can hardly help turning again to the quasi-phenomenal representation of the unity of the organism.

All the theories of the determination of form by agents which are in, but which are not, the protoplasm which we know, are merely commentaries upon a general feeling that this minute mass of unstable matter, the germ, cannot keep intact the image which is entrusted to it, and cannot contain the whole unity of the organism : and that, even if the germ, as we know it and without special agents, can determine one form with that extraordinary definiteness which we know, yet it cannot be the determination of two or more different forms, which appear to be adjusted to one another with an infinite complexity, in defiance of the necessities of their common origin with one another, from an homogeneous protoplasm.

For all these theories deny that any protoplasm which we know can be the unity of the organism, that is, can hold the image or the alternative images and no other, intelligibly within it. Yet popular teaching ordinarily proceeds upon the ground that there is a special protoplasm for every specific, and therefore for every individual form, and that the nature of this protoplasm is the complete determination of the individual form. Now we know enough to be able to contradict this statement, so long as it takes into account any of those qualities of protoplasm which are, or which may be known. But when it has recourse to qualities which we neither know nor can know, then we can only leave it alone as an hypothesis which is not aware even of its own difficulties. Let us take an example. The protoplasm of a somatic cell in a plant is necessarily very different from

THE UNITY OF THE ORGANISM. 203

the protoplasm in the ovule or pollen-grain of that plant, and the differences lie before the eye. There is, however, in that somatic cell, an identity with the germ cell, which is of such a nature that the somatic cell can take its place in reproducing the whole. Now, what is that identity? In every particular which is known to research that somatic cell is infinitely more like the somatic cells of other plants than it is like the germ cell of its own plant. And the identity is of such a nature that all these "superficial" and phenomenal differences are as nothing to it. And I think we have reason to believe that observation is likely, in its progress, rather to increase our knowledge of "superficial" differences between the germ and the somatic cells than to lead us to some very striking phenomenal identity between them. The presence of agents in each, to secure that identity which does not reveal itself in any particulars that we know, has, I think, been sufficiently dealt with. And that identity is certainly not abstract protoplasm.

This matter has been admirably studied by Dr. Stirling in "As Regards Protoplasm," but he leaves the reader in just such darkness as we are in, that darkness to which an unprejudiced study of the facts and the theories leads those who attempt to find an expression for the immanence of the whole in the part in the body of any creature. For the identity in difference is not phenomenal as a difference among others. It must not be quasi-phenomenally so represented. And yet, let it be what it will, it is the beginning and the end of this science of biology.

If we ask whether it is the unity of feeling, we are not necessarily recurring to the anthropomorphic conception of the immanent soul. For the latter is an agent which derives its only value from some sort of dim quasi-psychical ideas as to what ought to be done, and which holds an explicit scheme of the unity of the organism in its mind. The unity of feeling is, on the other hand, no more than a certain side of all changes which take place in an organism.

Indeed, it has been made into the sole definition of an organism. We need not question the probability of the fact that there is in every organism a unity of feeling. It would merely resolve itself into a discussion as to what we are to consider as an organism and what as a colony, and no hard line can be drawn upon that matter. Connection by filaments of protoplasm is now often demonstrated where before a complete wall was supposed to exist, and in any case we need hardly demand such protoplasmic continuity in order to the existence of the unity of feeling. Of course colonies, between the members of which no continuity exists, exhibit specific form, just as even bacterial cultures have each their distinctive appearances. And perhaps this fact alone should be enough to lead us to deny that feeling has a place by itself in the development and maintenance of the image. But it has so often been vaguely brought forward as in some way the secret of unity that its claims ought to be considered. Only, the mere fact that it is something of its own kind, and again, that it cannot easily be observed, gives it no special value in this connection.

The parts, changes, and experiences of an organism come together to form a certain emotional tone. Pain and pleasure—for these form the only distinction at which we have generally arrived—attend the manifestations of the creature's life. We may look on these as variations from the usual level of feeling, and we may suppose that level to have a certain distinctive character of its own, although it is not otherwise known than as departed from. But the fact that the normal emotional tone is not merely the negation of these—its manifestations, is sufficiently shown, not only by a more acute observation, which usually finds pleasure and pain, but also by the fact that its level may be altered by the stimulations which give feeling only by contrast to it. It is always there, though it is not recognised except as being changed. Its changes are recognised in proportion to their suddenness and rapidity, and it quickly weakens or eliminates them by rising or falling to the new level. We must ascribe to the unity of feeling a permanent and continuous existence. But it is not easy to express it otherwise than by the above clumsy metaphor of an ever-changing level, with distinguished variations above and below it.

Now, this unity of feeling is a thing by itself, in that there come into it all the parts and experiences of the organism. It is thus like an agent which understands all that is going on, only that in the latter case every difference in the organism has its place as an explicit difference in the mind of the agent, whereas in the case with which we have to do the resultant state of feeling is an apparently homogeneous unity, for which each change in the body has its value, but in which each loses its character as a separate thing. The unity of feeling has the additional advantage of being a fact, whereas the agent is not a fact.

Now, a mere unity without differences will not help us. But though homogeneous, as seen in its beginnings, this is yet not without differences. Though the latter do not live in it as separately recognisable elements, they yet open out from it in an orderly development. I need not reiterate that which everyone who is familiar with Hartmann's biological work knows already, how that author regards the processes of the formation of the individual, organic reflexes, habits and instincts, and even the psychologically implicit movements of argument, as forming one kind; and how he regards that kind as, at any rate most easily, interpretable in terms of feeling. Movement which takes hold on reason without the mediation of understanding, development of system of difference out of an idea which is no one's idea, the rise of an orderly manifold out of a unity which precedes it and is immanent in it—these all come nearest to him in the analogy of the development of our explicit thought and our explicit action out of the unity of feeling. The "clairvoyance of the unconscious" is nothing but the implicit reasonable articulation of that unity, and the ingenious argument which deals with it does little more than to extend analogies from the unity of organic life. For as reason without words instructs the creature and its parts to movements which are purposeful, but of which it cannot see the end, informing the building of the body, and all blind instincts, so, and with no hard line of division, it is made to seem in the end to come into light as the "truth within ourselves," which "takes no rise from outward things." I am not going to give a hurried exposition of Hartmann, or of the interesting developments from him which are given to us in Geddes and Thompson, and later in Drummond. Here it is enough if we recognise that in the unity of feeling we have for the first time come upon a unity which develops its differences out of itself, and which is yet a phenomenal factor in organic life.

The view derives its special interest from the fact that both human experience on the one hand, and the simplest reflex or habit or morphological structure on the other, are accompanied by a specific change in feeling, which seems to be of importance in every case. Undoubtedly, all the *tropisms* and *tactisms* of the newer and living embryology are nothing but terms for irritability; and irritability, because of its apparently unintelligible and unexpected results, is easily thought of as feeling. You may say "accompanied by feeling" if you like; it is a barren distinction, and, of course, they may, and even must, be susceptible of a chemical interpretation. It is familiar that the reactions of irritable parts are first remarkable as having degrees and directions which are answerable, not so much to the degrees and directions of the stimulus, as to the advantages of the organism. In this respect, the movements and the life changes of the simplest animals, no less than the earlier ontogenetic changes of the metazoa, lend themselves most readily to an interpretation in terms of feeling, the complete scheme of the life or the embryo being in some way present in the unity of feeling. The demonic power of feeling, as shown in the insistence of lowly creatures upon their normal activities, even under repeated hindrances, together with the always purposeful nature of its manifestations, lead naturally to the thought that here we have a principle which is capable of retaining intact a reasonable scheme of unity, or an image, throughout the most careless and cruel incursions of circumstance. The development of feeling into the actions which, even under unfamiliar conditions, are necessary for the life, and its forethought which is past the power of thinking, strengthen the impression that the unity of feeling is the side of an organism upon which it takes hold upon reason. Undoubtedly, it is urged, feeling is reasonable; and that it seems to pass without break or barrier, psychologically, into understanding, is made much of in Hartmann's argument.

In feeling, for the first time, we come, therefore, to the principle of the normal. This is the beginning and end of vitalism. To take the simplest instance, departure from the unity means pain, and the fulfilment of it means joy. Now, feeling is, in science, always associated with the normal and the abnormal, with the advantageous and the hurtful, but usually, as it happens, by all the stiff joints and cranks of the Darwinian mechanism. A certain experience is hurtful, and, therefore, by natural selection, or what other factor you please, it comes to be associated with pain. Always empty, that form of explanation becomes, in the matter of feeling, quite ridiculous. And we are left with the fact of an immediate and intelligible indication on the part of feeling as to what must be avoided and what must be approached. It would seem as though we had here a function of the organism which brings all other functions into the unity of its own state, and determines them into teleological activity, and suffices, formally, even for a sort of rational embryology or physiological morphology.

And objections which might be raised upon the ground that feeling has no power over the organism would be set down chiefly to inexperience. We might answer them by bringing forward, in the approved method, the very miracles in this kind. Stigmata, the determination of after and complementary visual images by suggestion, the dilation of vessels and the trophic influences of joy, and the poisoning of the whole body by hate, envy, and terror, and a thousand other remarkable effects might be accumulated. Introspection itself, with all its partiality and its sharp division as of mountains rising above mist, would furnish enough evidence to show that even the changes in feeling which are recognisable because of their sudden intensity appropriate the most profound physiological effects. We early read character on faces, too early to have learned it by mere association, and progress in the art is due rather to experience of ourselves than to exhaustive analysis of others. But the miracles, after all, are the least part of evidence. The simple reactions of unicellular creatures, and the activities of germinal and somatic cells, are, indeed, metaphysically referred to chemical and mechanical properties, and are rightly so referred. But they are never, in practice, described, except with the implication of associated and governing feeling. And since they are the movements of organic individuals, it is impossible to describe them otherwise. Activity upon occasion, and result upon cause, would, in the ordinary practice of science, be interchangeable terms, were it not that the former introduces, logically, the conception of individuality, and physiologically, the mediation of the unity of feeling. Certainly, this unity has, in respect of its tenacity of the scheme of life, and of the pattern of form, in contempt of accident, in respect of its unique unity as of an idea, into which differences come and from which they are created and appear, and in respect of its universality as a side of all organic changes and parts, a strong apparent right to be called that unity of the organism for which the theories seek.

But though we have in feeling an actual and efficient indication of necessity, and a factor of incomparable power in organic life, and a certain unity which receives and creates all the differences, yet we have not, in it, attained to that unity after which the theories necessarily seek. It is not the universal which is in all the parts, nor the unity in which all qualities and periods, whether existent, latent, or alternative, are made possible. It is not that which retains the image or alternative images, so as to educate them from the germ, or to restore them when they are mutilated. It is not that which preserves the form in the encysted ciliate. It will not, in fact, take the place of the anthropomorphic agent of the theories, though it has, at least, the advantage of not erring against the rights of physiology.

It is not adequate to this task because, in the end, itself is only a part or aspect of the organism, and itself is conditioned by the rest. What is to be painful and what is to be pleasurable is dependent, I suppose, upon the structure and condition which is already there. We do not escape from "mechanism," although we are half trying to do so. The normal is what is fitting to the rest, but the rest is given. Every part can still only be referred to the others in an endless round of reciprocity; and the mere fact that the parts are united in feeling does not warrant us in giving a special place to the later as custodian of the image. The stability and accuracy of the image, in spite of its passage through the minute and unstable germ, and of its working out through actual process in the complex ontogeny, is not at all explained by the fact that the normal is secured by feeling; for it is not secured as the normal, but as the necessary complement for what is already there. And in that primary germ character which reveals itself in the development of the individual, the quality of feeling is but one part, as much dependent upon the rest as the rest is upon it.

The inner identity of the germ and the somatic cell, the identity of male and female, that of one somite with another, and the inner identity of the embryonic tissue of regeneration with the specialised tissue which has been removed—these do not imply merely a manner of thinking or anyone's opinion. It is a familiar and essential fact and conception for all physiology and pathology, for zoology and botany. These differents are not similar up to a certain point or level, or identical in this and that particular, and then and separately different in those other particulars or above that level. We have not to do with animals or plants which are sexless in ninety-nine particulars (being identical in the sense of the absence of difference), and are sexually diverse in one particular (being different in the sense of the absence of identity). These, and all such differents, are, in the ordinary experience of research, wholly different and wholly identical. Nothing could be more unlike a specialised organ than the small round cells which regenerate it, yet we cannot express the process of regeneration, even apart from hypothesis, otherwise than by reference to an inner identity which is unaffected by the most profound and obvious difference. What, again, could be more absolutely removed from one another, within the bounds of unicellular form, than ovum and spermatozoid; yet, again, what two cells could be more absolutely identical? If biology has anything to explain and to face, it is this unity, this identity in difference. The reader's mind will probably recur at once, so fixed are these empty habits, to some supposed chemical identity, or identity of very fine structure which cannot be observed, or identity of protoplasm in some unknown respect. But one should succeed in eliminating all such half-formed metaphors from thought, otherwise it is impossible even to see the problem. Now, as all other hypothetical forms for the unity of the organism have broken down, so does this unity of feeling break down and become inadequate for this function.

Feeling serves very well to unite all the parts of the

body, and to secure purposeful reactions. An increasingly strong case is being made out by research in favour of giving it a high place as a factor of ontogeny. But it is not present, as a whole, in each of the parts, or, at least, it is not so present as for physiology. As the reactions of germ and somatic cells, or of the two germ cells, are different, so must their feeling in its quality and conditions be different. The unity of feeling may very effectually bring an isolated cell into a whole, which should direct its parts in a purposeful manner, but it is difficult to see what grounds we have for supposing that there is, present in the isolated cell, the idea for the whole body as implied in a particular quality of feeling. That is simply to say, that if you give a physiological place to feeling, it ceases to serve as the beginning out of which all the appearances come. The "unconscious" of Hartmann, which, in so far as it is applied to organic life, associates itself only with feeling, can only do what it does in organic life because it is apparently dissociated from conditions by a transparent and contradictory shift. For, at the points at which it is necessary that it should come in—and it only thus enters as a friend in need—it takes the whole matter into its own hands, and entirely supplants, but does not enter among, the conditions in relation to which changes are ordinarily explained. It is, in fact, the universal which comes in as one particular interfering with other particulars in the same scheme, and this is a method of explanation which we have found reason to utterly condemn.

There is, however, one expression which is left to us for the unity of the organism, but it does absolutely no more than give us a fitting term with some suggestions of analogy, and throws no light on the nature of the unity. *Character* is the identity in difference of concrete individuals, and is the familiar expression for the whole of a system. Though we have been unable to reduce all qualities and parts to the position of functions of one quality or part which should be the creature *par excellence*, serving as the unity under the manifold, and as the persistent through the changes, yet we can regard all the particulars as manifestations and components of one character. That character may develop itself in ontogeny, but it does not change. It is the same in the simplicity of the germ as in the complexity of the image. It is identical under the differences of male and female. It is the common nature, though no common quality, of germ and somatic cell, and of the elements of the different tissues. Individuals which differ from one another, differ by one difference, which, however, cannot be described except as an infinite number of differences, and all the features of one individual are one character. This is not the character of the protoplasm, nor of the idioplasm, nor of the immanent soul, but of the whole creature. And this character is no cause or condition amongst others. It is an aspect of all, and is that aspect by which all comes into unity.

Now, such a conception, it may be urged, is no conception for an observational science. There would be more reason in this objection than there is, if it were not that the conception is shadowed forth in so many self-contradictory forms such as those which we have studied. It has proved itself necessary to the science of biology by all the writings of Spencer, and of Naegeli, and of Galton, and of many other authors who might be named. We are forced to find some expression for it. But apart from that fact of its universality in the science, we may consider its necessity to a study which deals only with individuals. When we say that such and such qualities exist together in an individual, we are pointing out, not their mere concurrence, but their identity. The parts of an organised whole belong to its *character*. And if we set out to build theories of the origin of qualities in the growth of the individual, and of their existence by reason of their significance to the individual, and especially if we treat these qualities all together and in general, as the hypotheses do, it is obvious that the beginning and end of our labour is to find just this expression for character. No solution can be found by assuming, more or less arbitrarily, that this or that feature is the character, and we do not move a step forward by inventing unknowable agents to represent it. The question therefore comes to be, what is the character as distinguished from the characteristics?

I have set down this question, and studied the biological answers to it, as clearly as I can. The further consideration of it does not belong to this essay. For my present purpose is served if the impulse, and the fallacy, and the relation to research, of the biological systems, have become at all apparent in the argument.

THE END.

Printed by Cowan & Co., Limited, Perth.

www.ingramcontent.com/pod-product-compliance
Lightning Source LLC
Chambersburg PA
CBHW022018220426
43663CB00007B/1129